WEATHER
WARFARE

Positive use of HAARP re hurricanes

Tangshan 88 for hostile doublers - 90

U.S. & ENMOD - 93
useful AIR FORCE table - 94

*"1978 * and then it all changed" 96*

ENMOD - 96 — what it is ★ 102
LITRARIS - 103

QUOTE - 112

Metallic al balané salts Carnican 33

by

(6/14/13) ★
Start at pa. 310 : PATTERN
313

Jerry E. Smith

Nicholas Begich DVD 2008
Angels Still Don't Play This HAARP

WEATHER WARFARE

ADVENTURES UNLIMITED PRESS

Weather Warfare

Published by
Adventures Unlimited Press
Kempton, Illinois 60946 USA

www.adventuresunlimitedpress.com

ISBN 1-931882-20-60-6

Printed in the United States of America

Cover artwork by J. R. Schumacher

10 9 8 7 6 5 4 3 2 1

WEATHER WARFARE

by

Jerry E. Smith

Acknowledgements

I want to thank J. R. Schumacher for both a wonderful cover and for being my First Reader and primary editorial consultant. Also thanks go to Ed Grondine for critiquing early drafts and to Kathy Collins and "Eagle One-Eye Cody" for preefrooding. Very special thanks go to Bob Sonderfan, Juanita Cox, Wendy Hall and the Saturday night "eaters" group for material and emotional support, research and believing in me in general. Special thanks go (in no particular order) to Brian Holmes, Jon Hix, Bill Gallagher, Jaye Beldo, David Hatcher Childress, Jennifer Bolm, Hunt Mossell, and Marshall Smith and to the many who would prefer to remain unnamed. I wish I could thank by name the small army of researchers who provided material over the last too many years! I do appreciate your work, and hopefully so will the readers of this book!

TABLE OF CONTENTS

INTRODUCTION [i]
Science Beyond The Box—HAARP—Geophysical Warfare—
Chemtrails?—Environmental Modification

Chapter 1: THE RAIN-MAKERS [1]

**Chapter 2: WEATHER MODIFICATION IN THE
TWENTIETH CENTURY [19]**
Cloud Seeding—Other Technologies—Hurricanes: Eye Of The
Weather Controversy—*Project Cirrus—Project Stormfury—
Project Popeye*

Chapter 3: EARTHQUAKES ON DEMAND? [61]
Project Faultless—Electromagnetic Waves And Earthquakes—
Tesla's Earthquake Machine—Scalar Weapons—*Project Seal*

Chapter 4: ENMOD LEGISLATION [95]
The National Weather Modification Policy Act Of 1976—Space
Preservation Act Of 2001—Weather Modification Research
And Technology Transfer Authorization Act Of 2005—The
Convention on the Prohibition of Military or Any other Hostile
Use of Environmental Modification Techniques (ENMOD)—The
Law Of War—Limitations On A State's Rights In Conflicts—
Environmental Modifications Covered By ENMOD—Crime And
Punishment Under ENMOD—Limitations Of ENMOD—*Air
Force 2025: Owning The Weather*

Chapter 5: HAARP [157]
What Is HAARP?—HAARP In The Media—HAARP Science—
HAARP Timeline—Completing HAARP—HAARP Today—Space
Shuttle Columbia—HAARP And UFOs—The Future Of HAARP

Chapter 6:CHEMTRAILS? [221]
What Is A Contrail And How Does It Form?—What About Persistent Contrails, Are They Chemtrails?—But Didn't These Persistent Contrails Just Start Appearing Recently?—Contrails And Global Warming—The High Bypass Turbofan—Global Dimming—Aviation Smog—Contrails And Scientific Research—What Has NASA Found?—Contrails And 9/11

Chapter 7: GEOENGINEERING [267]
A Sunscreen For The Planet?—Dr. Strangelove—Stratospheric Welsbach Seeding—Patently Obvious—Alaska Flight 261— "Deep Shield"—Other Theories

Chapter 8: CONCLUSION [309]

ABOUT THE AUTHOR [327]

APPENDIX A [329]
CONVENTION ON THE PROHIBITION OF MILITARY OR ANY OTHER HOSTILE USE OF ENVIRONMENTAL MODIFICATION TECHNIQUES (complete text)

APPENDIX B [337]
HAARP Patents

APPENDIX C [341]
Weather Modification Patents

BIBLIOGRAPHY [351]

INTRODUCTION

There are some reports, for example, that some countries have been trying to construct something like an Ebola virus, and that would be a very dangerous phenomenon, to say the least. Alvin Toffler has written about this in terms of some scientists in their laboratories trying to devise certain types of pathogens that would be ethnic specific so that they could just eliminate certain ethnic groups and races; and others are designing some sort of engineering, some sort of insects that can destroy specific crops. Others are engaging even in an eco-type of terrorism whereby they can alter the climate, set off earthquakes, volcanoes remotely through the use of electromagnetic waves.

So there are plenty of ingenious minds out there that are at work finding ways in which they can wreak terror upon other nations. It's real, and that's the reason why we have to intensify our efforts.

~ William S. Cohen, U.S. Secretary of Defense, 1997

This startling statement was made by William S. Cohen while he served as the United States Secretary of Defense. It is from his keynote address to the Conference on Terrorism, Weapons of Mass Destruction, and U.S. Strategy, given on 28 April 1997 at the University of Georgia, in Athens, Georgia. It can therefore be taken as an official position of the United States government, at least at that time.

Zbigniew Kazimierz Brzezinski, a Polish-American

political scientist, geostrategist, and statesman who served as United States National Security Advisor to President Jimmy Carter from 1977 to 1981, wrote in his 1970 book *Between Two Ages*:

> It is ironic to recall that in 1878 Friedrich Engels, commenting on the Franco-Prussian War, proclaimed that "weapons used have reached such a stage of perfection that further progress which would have any revolutionizing influence is no longer possible." Not only have new weapons been developed but some of the basic concepts of geography and strategy have been fundamentally altered; space and weather control have replaced Suez or Gibraltar as key elements of strategy.

After events like the Christmas 2004 Asian tsunami and 2005's record-shattering Atlantic hurricane season many people have wondered just how "natural" those natural disasters were. Could terrorists or rogue states have been involved? Is it really possible to "alter the climate, set off earthquakes, volcanoes remotely through the use of electromagnetic waves"? Has "weather control" really become a key element of national strategy?

While the use of electromagnetic waves (radio frequency weapons) is exceedingly difficult to prove, there indeed are many out there, both on the drawing board and in the arsenal. Secretary Cohen wasn't just making them up as a ploy to get increased funding for counter-terrorism efforts. He knew about them, and so should we.

SCIENCE BEYOND THE BOX

"Everybody talks about the weather, but nobody does anything about it." Mark Twain [Samuel Langhorne Clemens (1835–1910)] was credited with saying that in an editorial by Charles D. Warner Dudley in the *Hartford Courant* of Hartford,

Connecticut, in 1897. As we will see in the next chapter, it wasn't entirely true then and is far less true today. This is a fine example of the limits of "conventional wisdom."

There certainly are many readily discernable civilian projects intended to do something about the weather today. These are taking place around the world as you read this. In the United States they include cloud seeding to increase the snow pack in the Sierra Nevada Mountains of California and hail suppression over farmland on the Great Plains, to name but two. We will briefly examine the positive and negative sides of civilian weather modification in this book, which may be just the "off the shelf" versions of military technologies.

Better than 80% of what I will present to you in this book is well established and fully cited mainstream science. But mainstream science has self-imposed blinders. Science and the base of knowledge it is founded upon resists change. There are many reasons for this. One is that in order to change, science (and individual scientists) must admit fallibility — something that most people of education are reluctant to do. Who wants to admit they were wrong? How much harder would it be to admit being wrong if advancing your career depended on your being right? Also, position within the scientific community (and grant money) does not go to mavericks. This institutionalized resistance to new theories has resulted in it routinely taking from 50 to 100 years for new discoveries to move from ridiculed "nonsense" to revered facts.

I bet you can think of several examples of this lag between the time a theory is presented and its universal acceptance. You may remember this one: Alfred Wegener, a German geophysicist and meteorologist proposed the theory of *continental drift* in his book *The Origins of the Continents and the Oceans*, published in German in 1915. For nearly 50 years his theory was ridiculed, yet today it is taught in nearly every school as the science of *plate tectonics*.

Today many branches of mainstream science have circled the wagons, waging war with "cranks" with goofy theories, some of which, like Alfred Wegener's, will eventually be proven correct. One such example is the battle between American archeologists and anthropologists and scores of "revisionist" researchers whose work collectively demonstrates that the Americas were reached many different times and in many different places by many more peoples than just a few Chinese crossing the Bering Strait land bridge 12,000 years ago.

Another example is the assault on traditional Egyptology by another army of independent investigators. While West and Schoch, whose research pushed back the age of the Sphinx by thousands of years, and Hancock and Bauval, who found a connection between the positions of the Giza pyramids and the constellation of Orion, have received the greater notoriety, the most spectacular work has been done by Christopher Dunn, a master craftsman and engineer. Dunn has virtually proven that the conventional wisdom of the ancients having used copper chisels and stone pounders to build the great monuments of Egypt is ludicrously wrong. His work clearly shows that the ancients used advanced machining techniques, especially high-speed lathes. Further, he has presented virtually irrefutable evidence that the Great Pyramid was not a burial chamber, but the largest machine ever built. So far the mainstream's only answer to him has been to ignore his work.

Similarly, understanding weather warfare will take us outside the box of conventional wisdom. Why is the weatherman so often wrong? Meteorology isn't a complete science. Many things about the atmosphere and its interactions with other geophysical systems, like the oceans and the earth's magnetic field, are still not fully understood. Meteorology, like all mainstream sciences, refuses to look at "crank" theories raised outside its circle. This includes anything that smacks of "astrology." Unfortunately this smugly arrogant stance has blinded meteorologists to important facts.

The single greatest influence on the Earth's weather is the Sun. It is an accepted fact that solar eruptions from sunspots affect the weather. The Sun's sunspot cycle seems to be affected by a number of factors, including the positions of the inner planets (because of their proximity) and Saturn and Jupiter (because of their size). The Sun's magnetic field is as large as the solar system and affects, and is affected by, everything within it. The solar system is just that, a system in which all pieces are interconnected. Understanding these relationships better should lead to improved forecasting — except it won't happen anytime soon, as meteorologists clap their hands over their ears and chant "I can't hear you" when you bring this up.

Temperatures across North America in January 2006 were record-setting. Where these temps natural or artificial? If manmade, were they inadvertently caused or was the US under eco-terrorist attack? Considering the topic of this book you might expect me to say that we were under attack — but I am going to offer an even more outlandish theory — these temps were perfectly natural and would have been predicted if only meteorology didn't have rectal-cranial inversion.

A "crank" theory that has been repeatedly championed in the popular press for the last few decades is the so-called "Jupiter Effect." This is the observation that some astronomical events have an effect on the Earth. Retired forester and fire fighter Roger L. Jewell spent a lifetime trying to discover if there was anything to the old adage about there being "fire weather" and if severe wild fires were the result of weather cycles. In his groundbreaking book *Riding The Wild Orb: Long-term Weather Extremes On The Planet* he outlines his discovery that there are indeed weather cycles caused by planetary mechanics, and not just the four seasons, which is about all the mainstream will admit to.

One of the things he discovered was that North America has a "January thaw," like the one we saw in 2006, every eight years like clockwork (this last one was the hottest because of other additional factors in play, like global warming). His discovery was that a January thaw happens every time the planet Venus comes between the Earth and the Sun around the time of the Winter Solstice. In astrology this is called a conjunction, and in meteorology it is called a crank theory because it sounds too much like astrology for these learned men to deal with. It is anyone's guess how much else the mainstream is responding to by squeezing its eyes shut.

I mention this in that this book is about human manipulation of natural forces to produce unnatural events — but how do you determine what's natural and what isn't? This is a difficult and delicate subject. As we will discuss later "Mother Nature gives great plausible deniability." One of the enticements of geophysical warfare is being able to hide it behind Mother Nature's skirts.

One needs a baseline of what is unquestionably "natural" to be able to say definitively that so-and-so is "unnatural." And thanks in part to the "half-built" nature of modern geosciences we really don't have that. This means that throughout this book I am going to have to tell you what is established and accepted by mainstream science, and then take you outside of the box. Rest assured I will always try to tell you when we cross the line from accepted to "crank."

HAARP

Getting back to Secretary Cohen's concerns about using electromagnetic waves to alter the climate, and/or set off earthquakes or volcanoes remotely…

One much discussed project that embodies both civilian and military geophysical applications is the High-frequency Active Auroral Research Program (HAARP). Although HAARP proponents claim it is nothing more than a simple civilian research station designed to investigate the properties of the upper atmosphere, few investigators buy that explanation.

HAARP does have the appearance of a civilian project with open access and the work being done by civilian scientists. However, the project is managed by a joint US Air Force and Navy committee and is funded out of the Department of Defense (DoD) budget. Most recently the heart of the program,

civilian etc.

the Ionospheric Research Instrument (IRI), was completed by one of the world's largest defense contractors working under the direction of the Defense Advanced Research Projects Agency (DARPA), the central research and development (R&D) organization for the DoD. DARPA manages and directs selected basic and applied R&D projects for the DoD pursuing research and technology "where risk and payoff are both very high and where success may provide dramatic advances for traditional military roles and missions."

Under construction since 1990, the HAARP IRI is a field of antennas on the ground in Southeastern Alaska. The facility was probably completed late in 2005 with the announcement of same added to the DARPA website in March of 2006. It is now the world's largest radio frequency (RF) broadcaster, with an effective radiated power of 3.6 million watts — over 72,000 times more powerful than the largest single AM radio station in the United States (50,000 watts). The IRI uses a unique patented ability to focus the RF energy generated by the field, injecting it into a spot at the very top of the atmosphere in a region called the ionosphere. This heats the thin atmosphere of the ionospheric region by several thousand degrees. HAARP, then, is a type of device called an ionospheric heater. This heating allows scientists to do a number of things with the ionosphere. Controlling and directing the processes and forces of the ionosphere is called "ionospheric enhancement." An early HAARP document stated:

> The heart of the program will be the development of a unique ionospheric heating capability to conduct the pioneering experiments required to adequately assess the potential for exploiting ionospheric enhancement technology for DoD purposes.

What might those DoD purposes be? Something about winning wars, eh? How might those purposes be achieved? What technologies will be needed to win the wars of the future? Researchers trying to answer those questions have come up with a vast number of possibilities, most bordering on science fiction. But then again, good science fiction is about recognizing the problems of the future, and suggesting solutions to them before they happen.

On 23 March 1983 President Ronald Reagan called upon "… the scientific community in our country, those who gave us nuclear weapons, to turn their great talents now to the cause of mankind and world peace, to give us the means of rendering these nuclear weapons impotent and obsolete." This quest for the creation of a technology, of a weapon or weapons system that would make atomic war impossible was officially named the Strategic Defense Initiative (SDI). The press lost no time in dubbing it *Star Wars* after the George Lucas movie.

That Initiative sent the United States military-industrial-academic complex on the greatest and costliest weapons hunt in human history. Thousands of ideas were floated, hundreds of those were funded. While SDI research has since been officially abandoned, some ideas are still being actively pursued to this day.

Not all of these ongoing developmental programs are taking place in laboratories of the military and its contractors. Some of these ideas involve technologies or applications that, as weapons, violate international treaties; others, the use of which would be repugnant to the ethical and moral values of the majority of Americans. In an effort to avoid public outcry (and international condemnation) some of these programs have been disguised as civilian science. One of those may be HAARP.

As Dr. Bernard Eastlund, the putative inventor of HAARP put it: "The boundary between science fiction and science comes with can you actually make the thing that you're proposing." Bernard J. Eastlund is a physicist who received his B.S. in physics from MIT and his Ph.D. in physics from Columbia University. He led a team of scientists and engineers working for Advanced Power Technologies, Inc. (APTI), a wholly owned subsidiary of ARCO. Eastlund's team developed the concept of a massive antenna array that could produce the kind of shield called for by President Reagan.

The APTI patents that HAARP is probably based on openly discuss manipulating the weather by moving the jet stream and using other techniques to create floods and droughts at will. These patents also describe a way to raise the ionosphere, sending it out into space as an electrically charged plasma capable of destroying anything electronic (like an incoming ICBM or a spy satellite) passing through it. HAARP certainly looks like a ground-based Star Wars weapons system, a "relic" of the Cold War. But unlike most such relics this one is up and running and now fully funded.

In August of 2002 the Russian State Duma (their version of Congress or Parliament) expressed concern about HAARP, calling it a program to develop "a qualitatively new type of weapon." A joint commission of the State Duma's International Affairs and Defense Committees issued a report that said:

> Under the High Frequency Active Auroral Research Program (HAARP) the USA is creating new integral geophysical weapons that may influence the near-Earth medium with high-frequency radio waves. The significance of this qualitative leap could be compared to the transition from cold steel to firearms, or from conventional weapons to nuclear weapons. This new type of weapon differs from previous types in that the near-Earth medium becomes at once an object of direct influence and its component.

The report further claimed that the USA's plan to carry out large-scale scientific experiments under the HAARP program, and not controlled by the global community, would create weapons capable of jamming radio communications, disrupting equipment installed on spaceships and rockets, provoke serious accidents in electricity networks and in oil and gas pipelines and have a negative impact on the mental health of people populating entire regions.

An appeal, signed by 90 deputies, demanding that an international ban be put on such large-scale geophysical experiments was sent to President Vladimir Putin, to the United Nations (UN) and other international organizations, to the parliaments and leaders of the UN member countries, to the scientific public and to mass media outlets.

HAARP technology, real or imagined, has shook up a lot of people, setting off alarm signals with environmental, health, political and sociological concerns in the minds of scientists, reporters and concerned citizens all over the globe. We will take up HAARP in detail later in this book bringing you up to date on what is happening there now.

GEOPHYSICAL WARFARE

Those who fight wars have known since the beginning of the human race that weather affects the outcome of battle. Since the Vietnam War the United States military has been actively trying to influence the outcome of conflicts by developing technologies that manipulate the weather. It is hardly surprising then that other nations, such as the former Soviet Union, China and even North Korea have been accused of doing the same. How far might the world's military have come in the decades since Vietnam?

This is the 21st century — we are living in the future! If you are over forty, many things that were science fiction when you were in school are now everyday realities. Control of the weather, at least to some degree, is an established and expanding field of scientific and commercial endeavor. But, is it really possible to create floods or droughts, steer hurricanes or set off tsunamis, much less trigger earthquakes or volcanic eruptions? Are these still in the realm of science fiction, or have they become science fact? Throughout this book we will examine evidence that intentional hostile control of the weather and other environmental processes (collectively called *geophysical warfare*) is a reality just as Defense Secretary Cohen and the Russian State Duma said.

In 1966 Dr. Gordon J. F. MacDonald wrote: "The key to geophysical warfare is the identification of environmental instabilities to which the addition of a small amount of energy would release vastly greater amounts of energy." This was in "Geophysical Warfare: How to Wreck the Environment," a chapter he contributed to Nigel Calder's 1968 book, *Unless Peace Comes: A Scientific Forecast of New Weapons*.

Dr. MacDonald was no off-the-deep-end loony; he was one of the United States' top internationally recognized scientists. At the time he penned those words he was Associate Director of the Institute of Geophysics and Planetary Physics at the University of California, Los Angeles (UCLA). [A side note: the UCLA Plasma Physics Laboratory (PPL) has, over the last twenty years, operated another ionospheric heater near Fairbanks, Alaska, known as HIPAS (for HIgh Power Auroral Stimulation), the success of which led to the creation of HAARP.] Dr. MacDonald was also a member of the President's Science Advisory Committee, and was later made a member of the President's Council on Environmental Quality.

In "Space" an article for *Toward the Year 2018*, released in 1968, Dr. MacDonald elaborated on the possibilities of geophysical warfare writing:

> By the year 2018, technology will make available to the leaders of the major nations a variety of techniques for conducting secret warfare, of which only a bare minimum of the security forces need be appraised. One nation may attack a competitor covertly by bacteriological means, thoroughly weakening the population (though with a minimum of fatalities) before taking over with its own overt armed forces. Alternatively, techniques of weather modification could be employed to produce prolonged periods of drought or storm, thereby weakening a nation's capacity and forcing it to accept the demands of the competitor.

In *Between Two Ages* Brzezinski openly discussed conducting war covertly by causing droughts, storms, volcanic eruptions and floods. If, as Dr. MacDonald wrote, only a few in the military or scientific community need know what is actually taking place, how would the general public know if such a war were being fought? Were the dead and displaced of New Orleans victims of an invisible battle? Many people around the world are convinced that a covert weather war has in fact been raging for years, possibly decades — the evidence for which we will examine in this book.

Dr. MacDonald was also a member of the Board of Directors of the MITRE Corporation, a not-for-profit federally funded R&D company. It was MITRE who did the outrageously flawed environmental impact study for HAARP that, after carefully ignoring where the radio waves went after leaving the antenna array concluded that there would be no impact on the environment!

He was also a member of two organizations that have been accused of engaging in globalist conspiracies by anti-New World Order researchers: the Council on Foreign Relations (CFR) and the secretive JASON Group. The CFR, with such related organizations as the Trilateral Commission and the World Federalist Association, is a prominent supporter of globalization and of global governance schemes, as can be read in their own magazine *Foreign Affairs*.

The JASONs are described as a "select group of world class scientists" who conduct studies for different parts of the U.S. government. The group is referred to as the JASON Defense Advisory Group, or simply the JASON Group. Today their headquarters is located in the JASON Program Office at the MITRE Corporation. JASON was created as an elite division within the Institute for Defense Analyses (IDA). The official history of IDA says:

> IDA traces its roots to 1947, when Secretary of Defense James Forrestal established the Weapons Systems Evaluation Group (WSEG) to provide technical

analyses of weapons systems and programs. In the mid-1950s, the Secretary of Defense and the Chairman of the Joint Chiefs of Staff asked the Massachusetts Institute of Technology to form a civilian, nonprofit research institute. The Institute would operate under the auspices of a university consortium to attract highly qualified scientists to assist WSEG in addressing the nation's most challenging security problems.

Over the years, IDA has modified its structure to remain responsive to sponsor needs. In 1958, at the request of the Secretary of Defense, IDA established a division to support the newly created Advanced Research Projects Agency. Shortly thereafter, the mandate of this division was broadened to include scientific and technical studies for all offices of the Director of Defense, Research and Engineering. Subsequent divisions were established to provide cost analyses, computer software and engineering, strategy and force assessments, and operational test and evaluation.

[Another side note: the Office of the Director of Defense, Research and Engineering (DDR&E) was also involved in the early planning of HAARP. Another major player in HAARP is DARPA, formerly the Advanced Research Projects Agency (ARPA), who created the ARPANET which eventually grew into the Internet. ARPA was renamed DARPA on 23 March 1972, then back to ARPA on 22 February 1993, and then back to DARPA again on 11 March 1996.]

Dr. MacDonald published numerous papers and articles on future weapons technology. These suggested such coming "advances" as manipulation or control over the weather and climate, including destructive use of ocean waves and melting or destabilizing of the polar ice caps; intentional ozone depletion; triggering earthquakes; and control of the human brain by utilizing the earth's energy fields. Today the polar ice caps are indeed melting and holes in the ozone layer are growing. Could these be the handiwork of advanced weapons? What about earthquakes and mind control? Are we, the private citizens of the world, in the crosshairs of bizarre, unthinkable weapons?

What about the Russian Duma's claim that HAARP could have a negative impact on the mental health of people populating entire regions of the globe? In "Vandalism In The Sky?" their seminal article on HAARP in *Nexus Magazine* Dr. Nick Begich and Jeane Manning describe how HAARP could be used to induce mental dysfunction, quoting from Brzezinski on a proposal from Dr. Macdonald saying:

> Political strategists are tempted to exploit research on the brain and human behavior. Geophysicist Gordon J. F. MacDonald—specialist in problems of warfare—says [an] accurately-timed, artificially-excited electronic stroke "...could lead to a pattern of oscillations that produce relatively high power levels over certain regions of the Earth... In this way, one could develop a system that would seriously impair the brain performance of very large populations in selected regions over an extended period..."

In 1969 Dr. MacDonald wrote: "Our understanding of basic environmental science and technology is primitive, but still more primitive are our notions of the proper political forms and procedures to deal with the consequences of modification."

It would appear that the gap between our understanding of environmental science and technology and our ability to grapple with this knowledge as a body politic has changed little in the intervening decades. This book, like so many others since Rachel Carson's *Silent Spring* is an attempt to correct this problem of science creating ever more and bigger monsters to come out of science's Pandora's Box, by alerting you to the need to find a solution. It is not sufficient merely that I, or anyone, write; it requires you, the reader, the citizen, to take action, to create the necessary "forms and procedures to deal with the consequences of modification."

CHEMTRAILS?

This question of whether advanced technology is being intentionally used to alter the climate is also seen in the contrail vs. chemtrail debate, which we will address in depth. If you have spent much time looking up lately you may have noticed that something about the sky seems to have changed. You may be old enough to remember when the condensation trails (contrails) from jet aircraft used to be thin streaks behind the planes, blazing like comets, disappearing in seconds. Now they persist for hours. The brilliant blue sky of the morning becomes a milky white opalescence in late afternoon as the contrails merge together, creating a layer of artificial cirrus clouds. What's going on? Are we being sprayed? Is there some chemical in these persistent contrails making them *chemtrails*? Who could be doing this, and why?

If you try to answer any of the questions connected to chemtrails via the Internet you will find many sites devoted to this subject, both pro and con. Some will tell you there is no spraying, that these strangely persistent contrails are in fact perfectly natural. Others will detail bizarre things found in the air. Chemicals and metals like aluminum and barium salts, and stranger things like E-coli bacteria and genetically modified human blood, are but a few of the bizarre things claimed to have been spewed out after commercial aircraft or mysterious unmarked tankers have unloaded their covert payloads over the heads of frightened or angry witnesses.

Particulate matter in the air is called an aerosol. If you search the scientific media for information on aerosols you will discover that there are literally scores of organizations (military, academic, commercial, environmental and governmental) involved in studying, monitoring, tracking and *placing* aerosols in the atmosphere. A number of technologies using aerosols for a variety of purposes have been patented, and many more are under development. Chemtrails, it seems, are not a "one-size-fits-all" phenomena. There is not just one program running or just one purpose to this alleged spraying. I have found that there are many players, many technologies for many purposes, injecting an amazing variety of aerosols into our environment.

In the chemtrails section of this book I will try to lay out for you what is real and what is probably lies and hysteria. Like in the UFO field, military intelligence may well have injected wild and crazy stories into the chemtrail debate to hide covert operations. Unfortunately too, the scientifically challenged have added a tone of very literal Chicken Little hysteria to all this. It is probable that not all contrails are chemtrails, and possible that not all chemtrails are intentionally evil—but all *are* an environmental hazard that will have to be addressed. Aerosols are a very real problem, somewhat acknowledged in the scientific press and totally ignored by the popular media.

After the week of no commercial flights over the United States in the wake of the horrific events of 11 September 2001 atmospheric scientists were able to take new measurements of an unmodified sky. They found that these "clouds" produced by persistent contrails—chemtrails if you will—did in fact reflect back into space solar energy during daylight hours and trap in heat at night, adding about .5 degrees Fahrenheit to the average temperature of North America.

In the contrail/chemtrail debate we see a bridge between the topics of hostile (military) use of environmental modification (EnMod) and the unintentional. It is now a fact recognized by the scientific community that the banks of artificial cirrus clouds left behind by the thousands of aircraft that ply our skies daily are taking a toll on the environment. Officially this comes under the heading of "inadvertent environmental modification" and is a part of the Global Warming debate. To date the most commonly talked about aspect of "inadvertent" environmental modification has been the destruction of the rainforest. But in just the last few years contrails and the "global dimming" they seem to be causing is being pushed to center stage in scientific circles. But what if contrails are actually *chemtrails*? What if these strangely persistent contrails are actually the product of some nefarious covert spraying operation(s)? We will look at several possible answers including the possibilities of misguided civilian programs as well as military operations.

ENVIRONMENTAL MODIFICATION

Examples of both hostile (military) and inadvertent modification of the environment can be seen back to the dawn of civilization. Sumeria is the oldest civilization recognized by Western science. The city-state of Sumer (Sumeria) existed between the Tigress and Euphrates Rivers in an area called The Fertile Crescent in what is now Iraq between seven and ten thousand years ago. Today that land is a vast region of deserts caused by over-farming and over grazing. Desertification is a chilling example of inadvertent environmental modification, a process that is ongoing in several parts of the world today. The United States got a taste of it in the "dust bowl" years of the Great Depression. Scientists are warning us that it could happen there again.

One of the oldest examples of hostile environmental modification (EnMod) can be seen in the salting of the fields of Carthage by Rome. After a long and costly war between those two great powers Rome won, and intent on never having to deal with Carthage again, sowed salt into the croplands of Carthage, sterilizing the earth and forcing the people to move away, abandoning their city so as not to starve.

Several examples of EnMod from American history can be cited. During the 1840s the United States government fought a series of wars against the indigenous peoples of North America. These were called the Indian Wars. The First Peoples (Indians) of the Great Plains had an economy and a way of life based on the buffalo. The US government realized that if the buffalo were gone the Indian would be destroyed. Thousands of buffalo hunters were hired and sent to slaughter the buffalo in the millions. The buffalo was nearly driven to extinction and the Indian was brought to his knees.

One hundred years later America's military pursued a similar course in Southeast Asia when fighting the Vietnam War. There a two-pronged attack was launched on the environment. In an echo of the tactics used against the Indians, military planners realized that the jungle was home to the enemy. Millions of gallons of defoliants, Agent Orange, Agent White and others, were poured over the jungles of Southeast Asia. This was not merely to clear paths and roads and landing fields, but was actually intended to destroy the jungle itself and the shelter and sustenance it provided to the enemy. Simultaneously an effort was made to make the weather a weapon. Under *Operation Popeye* cloud seeding was conducted over the Viet Cong's resupply routes, collectively called The Ho Chi Min Trail, in hopes of inducing drenching rains, turning these dirt tracks into impassible quagmires of mud.

While the EnMod practiced during the Indian Wars was a heartbreaking success, the tactics used during the Vietnam War proved to be as unsuccessful and destructive as the war itself. The horror and folly of military EnMod led to the creation and passage of the United Nations sponsored treaty named The Convention on the Prohibition of Military or Any Other Hostile Use of Environmental Modification Techniques (widely known as the EnMod Convention or simply ENMOD for short). It attempts to prohibit using the environment as a weapon in conflicts between nations.

After being co-sponsored by the Unites States and the Soviet Union it was formally adopted by the United Nations General Assembly on 10 December 1976 and officially opened for signature on 18 May 1977. ENMOD entered into force when Laos, the twentieth State Party, deposited its instrument of ratification on 5 October 1978. To date the EnMod Convention has only been ratified by 70 of the 193 countries recognized by the United Nations. Even worse, ENMOD is unenforceable in any practical sense. We will examine this in depth also. Meanwhile the US military has not lost interest in environmental modification, even if it is forbidden by the UN.

The Chief of Staff of the United States Air Force, General Ronald R. Fogleman, tasked the Air University at Maxwell AFB, AL to look 30 years into the future to identify the concepts, capabilities and technologies the United States will require to remain the dominant air and space force in the 21st century.

The Air University commander led a team of students and faculty from the Air University's Air War College and Air Command and Staff College; scientists and technologists from the Air Force Institute of Technology, located at Wright-Patterson AFB, OH; Air Force Academy and AFROTC cadets from around the country; and selected academic and business leaders in the civilian community across the nation in the 10-month effort to meet General Foglemans tasking.

The resulting study is called Air Force 2025 or 2025 for short. The team's findings were briefed to General Fogleman in June 1996 and to the Secretary of the Air Force, Dr. Sheila Widnall, in July 1996. The 2025 study was subsequently published in a collection of white papers consisting of an executive summary and 41 individual papers, totaling more than 3,300 pages of text.

One of those white papers was *Weather as a Force Multiplier: Owning the Weather in 2025.* In it we see what some military planners are thinking for future environmental modification applications, as we will examine in detail later. That paper began with:

In 2025, US aerospace forces can 'own the weather' by capitalizing on emerging technologies and focusing development of those technologies to war-fighting applications. Such a capability offers the war fighter tools to shape the battlespace in ways never before possible. It provides opportunities to impact operations across the full spectrum of conflict and is pertinent to all possible futures. The purpose of this paper is to outline a strategy for the use of a future weather-modification system to achieve military objectives.

While the *2025* study is not official US policy, it was drafted with the intent that it would provide a platform to build policy (and weapons systems) on. In this book we will look at several technologies under development and the policies and agendas those technologies could be used to further.

Before we look at the ways the world's militaries are seeking to own the weather today, let us step back and take a look at the development of this technology from the beginning of the modern scientific age.

CHAPTER ONE

THE RAIN MAKERS

Weather modification isn't just about giant storm systems. Rainmaking and hail prevention are considered established arts in some countries. China has 35,000 people engaged in weather management, and it spends $40 million a year on alleviating droughts or stemming hail that would damage crops. Russian officials claim to order up clear skies for Moscow's May Day parade. It's done by saturating clouds with dry ice, producing so many tiny droplets that drops can't grow big enough to fall as rain -- at least for a while.

In the U.S., though, there is no clear consensus on how well such techniques work, or if they work at all. In the 1970s the U.S. plowed $20 million a year into cloud-seeding research, but almost all federal funding has since dried up.

Nevertheless, dozens of state, local, and private operations continue in 10 states, including California, Idaho, Nevada, and Utah. Vail Mountain in Colorado and many other ski resorts pay for cloud seeding, and Vail estimates that teasing more precipitation from clouds boosts its snowpack by 15%.

So reported Otis Port in "Rainmaking Has Its True Believers—And Skeptics" in *Business Week* on 24 October 2005. Human intervention in and possible control of the weather and other environmental processes are not fairy tales or wishful thinking, but at least to some degree are off-the-shelf technologies in the 21st century—people are indeed

1

doing something about the weather.

Mankind has always had a keen interest in the weather. Throughout human history we have seen the effects of weather on crops, and the loss of life and property through the violence of storms. In ancient times people made sacrifice to the gods in a crude attempt at influencing the weather. In many parts of the world people still conduct elaborate rituals for rain and fertility—the rain dances of aboriginal First Peoples (Indians) in the American West is an example.

Mainstream science today admits that humans modify the weather both deliberately and unintentionally. The American Meteorological Society has at last grudgingly admitted that: "Evidence accumulated over the last 40 years suggests that certain local weather conditions including fogs, low clouds, and precipitation in some areas can be altered by carefully controlled cloud seeding." Similarly, today scientists and citizens alike better understand the effects of inadvertent weather modification. Cities and industrial complexes do affect local weather conditions and alter precipitation. Regional weather changes result from other human activities such as deforestation and vehicle traffic on major transportation corridors.

As I mentioned in the Introduction to this book, the hostile use of environmental modification can be traced back to ancient times, such as in the salting of the fields of Carthage by Rome in 146 B.C. From relatively modern times I cited the intentional destruction of the buffalo by the U.S. during the Indian Wars and more recently the millions of gallons of defoliants poured over the jungles of Southeast Asia concurrent with cloud seeding over the Ho Chi Min Trail during the Vietnam War.

To understand where this technology stands now, and to get an idea of what's coming in the future, we need some knowledge of its history. A line drawn from the beginnings of this technology through its development to the present could point towards a foreseeable future.

The modern interest in making rain for profit and/or

the public good began following the American Civil War. A surprisingly large volume of literature on the subject was generated between 1890 and 1894 alone. Martha B. Caldwell in her article "Some Kansas Rain Makers," published in the *Kansas Historical Quarterly* in August of 1938 summed up much of this material. She wrote:

> These writers had various theories as to the methods of producing rain. A French author suggested using a kite to obtain electrical connections with the clouds. James P. Espy, a meteorologist from Pennsylvania, proposed the method of making rain by means of fires. This idea is prevalent on the Western Plains where the saying, "A very large prairie fire will cause rain," has almost become a proverb. The Indians on the plains of South America were accustomed to setting fire to the prairies when they wanted rain. A third method patented by Louis Gathman in 1891 was based on the supposition that sudden chilling of the upper atmosphere by releasing compressed gases would cause rapid evaporation and thus produce rain. One of the oldest theories of producing artificial rain is known as the concussion theory, or that of generating moisture by great explosions. The idea originated from the supposition that heavy rains follow great battles. Gen. Daniel Ruggles of Fredericksburg, Va., obtained a patent on the concussion theory in 1880, and urged congress to appropriate funds for testing it.
>
> By 1890 the subject of artificial rain making had attained considerable dignity; two patents had been issued and through the efforts of Sen. C. B. Farwell, Congress had made appropriations, $2,000 first, and then $7,000 to carry on experiments. In 1892 an additional appropriation of $10,000 was made to continue the work. The carrying out of these experiments naturally fell to the Department of Agriculture, and the Secretary selected R. G. Dryenforth to conduct them. In 1891

Mr. Dryenforth with his assistants proceeded to the "Staked Plains of Texas" to begin work. Included in the equipment which he took with him were sixty-eight explosive balloons, three large balloons for making ascensions, and material for making one hundred cloth covered kites, besides the necessary explosives, etc. He used the explosives both on the ground and in the air. An observer stated that "it was a beautiful imitation of a battle." The balloons filled with gas were exploded high in the atmosphere. After a series of experiments carried on in different parts of Texas over a period of two years, his conclusions were to the effect that under favorable conditions precipitation may be caused by concussion, and that under unfavorable conditions "storm conditions may be generated and rain be induced, there being, however, a wasteful expenditure of both time and material in overcoming unfavorable conditions."

These government-funded tests were much talked about and helped to make people conscious of the possibilities of controlling the weather or at least making rain. Individuals throughout the United States began to conduct experiments based on these and other theories. Desire to perfect this technology was driven to a fever pitch by a drought that began in 1891 and lasted for several years, affecting most of the states and territories west of the Mississippi.

Conditions in Kansas in the early part of 1891 had initially been favorable, with rain falling in sufficient quantities to mature the early crops; but by the end of July the drought had set in and corn and other grains began to wither under a scorching sun and incessant hot winds. "The farmers," Ms. Caldwell wrote, "in their helpless condition were ready to grasp at the last straw, which in this case happened to be the Rain Maker."

The fame of a certain Frank Melbourne, said to be an Australian, as a "rain wizard" had been spreading throughout the country. She tells us:

> Marvelous stories were told of his operations at
> Canton, Ohio, where he was said to so control the
> weather that he could "bring rain at a given hour."
> Since he was fond of outdoor sports he "so adjusted
> his machine that all the Sunday rains come late in the
> afternoon, after the baseball games and horse races
> for the day are over." Mr. Melbourne said his machine
> was "so simple that were its character known to the
> public every man would soon own one and bring rain
> whenever he felt like it." The editor of the Hutchinson
> News thought there would be serious objections to
> this for "there could never be a political barbecue
> without all the rain machines of the opposition being
> set in motion," and the "infidels would spoil all the
> camp-meetings and the church people ruin the horse
> races."

Mr. Melbourne was contacted about the first of September
in 1891 by Mr. A. B. Montgomery, of Goodland, Kansas, in the
hope of bringing him, and rain to Goodland. The Rain Maker
set his price for a good rain at five hundred dollars, and
stipulated that his rain would reach from fifty to one hundred
miles in all directions from the place where he operated his
equipment. Ms. Caldwell tells us:

> A meeting to consider the matter of making a
> contract was held at the courthouse. It was apparently
> an enthusiastic one with a large crowd in attendance.
> Two committees were appointed, one to contract with
> Mr. Melbourne and another to make arrangements
> for the occasion. A considerable sum was raised at the
> meeting and the citizens were admonished not to shirk
> their duty in the effort to have Mr. Melbourne there on
> or about September 25. "Let every farmer who is able
> act promptly and contribute to this fund," advised

the Goodland News, "and we will give to Goodland and Sherman County a valid boom such as they have never enjoyed before."

The purse of five hundred dollars having been raised, the papers announced that Melbourne would be in Goodland the 25th. Plans were made for a great occasion. The county fair was to continue over Saturday. People were expected from all over the country, Gov. Lyman Humphrey and his staff, and Sec. Martin Mohler and members of the State Board of Agriculture were to be Goodland's guests. Saturday was to be the eventful day with horse racing, speaking and other entertainment culminating in a grand ball in the evening.

A two-story building twelve by fourteen feet and fourteen feet high was built on the fair grounds for the operations of the Rain Maker. The upper story, containing four windows facing the different points of the compass, was Melbourne's work room. The room also contained a hole in the roof four inches in diameter for the escape of rain making gases. The lower room was used by Melbourne's brother and his manager who served as sort of a body guard to the Rain Maker.

Not everyone was onboard with the enterprise. One man in Goodland said that he would not give anything because rain making "was interfering with the Lord's business and harm would come out of it." Another declared he did not believe in it and "the first thing we knew we would have a hell of a tornado here that would blow the town from the face of the earth." Still another called it a humbug (a hoax or fraud—yes, it is a real word, not just a noise uttered by Ebenezer Scrooge) "because the clouds were beyond the reach of man and controlled by the Lord and when man went to tampering with them he was setting himself up against the divine powers."

Such attitudes continue to this day. Even in the 21st century prominent and famous people have expressed their opinion that bad weather was not merely a random chance of nature but directed by divine or supernatural powers. Hurricane Rita hit New Orleans just days before an annual gathering called "Southern Decadence," a week-long party by and for gays and lesbians, was set to begin. Televangelist Pat Robertson and the city's own Mayor, C. Ray Nagin, both suggested that Rita was "divine retribution." For these folks the phrase "Act of God" has a very literal meaning.

Many New Age writers have expressed the belief that Gaia, the Earth Goddess, is intentionally creating bad weather, earthquakes and such as a way to "balance" the earth. The Gaia Hypothesis proposes that our planet is alive and functions as a single organism that maintains the conditions necessary for her own survival. The Gaia Hypothesis was formulated by James Lovelock in the mid-1960s and published in his book *Gaia: A New Look at Life on Earth* in 1979. This book's reading title, *The Military's Plan To Draft Mother Nature*, is a somewhat tongue-in-cheek reference to The Gaia Hypothesis, suggesting that the Earth Goddess, like the old taunt about one's mother, may now be wearing combat boots. But I digress…

Ms. Caldwell continues:

> Mr. Melbourne, with his brother and manager, F. H. Jones, arrived on Saturday, the 26th, and were met at the station by the committee and a crowd of curious people. Much to his dismay light showers fell on the 25th and 26th and it was decided to postpone operations until Tuesday, during which time his expenses of ten dollars a day were paid by the committee. This was considered necessary to keep him from going to Topeka in answer to a call. Again on Tuesday night a light shower fell, but on Wednesday he took his rain apparatus to the fair grounds to begin work. He performed his work in great secrecy; no one was allowed within the building, and to keep the inquisitive from coming too close a rope barrier was erected about twenty feet from the building and the windows were curtained. However,

everyone went up and "gazed" at the building and the small hole in the roof through which cloud making substances escaped. "It was no more than looking at any frame shed," wrote one, "but to know that inside a man was dealing in the mysterious, made the place a curiosity."

Unfortunately for Melbourne his Wizard-of-Oz-like production initially failed to produce anything but clouds which were soon driven away by the wind. For two days the wind kept blowing at a steady thirty to forty miles per hour from the southwest, driving his gases to the northeast. Amazingly, heavy rains did fall in the northeast! Indeed, they were reported to be heavier than had ever been known at that time of the year! According to the Goodland *News*, large numbers of telegrams and letters were received begging for Melbourne's apparatus to be shut off. The committee in charge dutifully stopped him for a day and a half. He resumed work for another two or three days but ended with only one light shower to show for his efforts. In spite of Melbourne's failure many of the citizens of Goodland still had faith in him. They accepted his alibi that cool nights and heavy winds were not conducive to rain making.

Shortly after Melbourne's departure Goodland was buzzing with the news that a company had been formed locally to carry on Melbourne's work. The Goodland *News* reported that "after much argument and work a contract was entered into between Mr. Melbourne and the company, whereby the company was to be told the secret, furnished with a machine and allowed to operate in any part of the country." The name chosen for the new business was the Inter-State Artificial Rain Company. Its business plan called for western Kansas to be divided into districts and for a certain fee it would supply each district with the amount of rain needed for the growth and maturing of crops. A central station was to be established from which "rain making squads" would be sent out when needed. By spring they expected to have all preparations

8

made and be ready "to furnish rain to the farmers while they wait."

The company began a series of operations in Oklahoma, Texas, and places as far away as Tulare, California, with successes announced in most trials. One of the company officers wrote another saying "I tell you, Marve, we have got the world by the horns with a down-hill pull and can all wear diamonds pretty soon. We can water all creation and have some to spare."

On their return home the Goodland *News* glowingly wrote:

> It is a happy hour for Goodland to know that she is not only the Mecca of the home seeker; the innermost chamber of these broad plains; the morning star among a hundred towns of western Kansas, but also that she holds within her grasp the scepter that even sways the clouds. It's a happy hour to know that we have but to smite the rock (a la Moses) and the water cometh forth. We are the people.
>
> The Rain Makers — E. F. Murphy and O. H. Smith — have returned from California and bring with them not only assured success, but much California gold.

Ms. Caldwell continues:

> The reported success of the Inter-State Artificial Rain Company inspired others to enter the field, and early in 1892 two other rain companies were organized at Goodland. The Swisher Rain Company of Goodland was chartered January 13, with a capital stock of $100,000. Dr. W. B. Swisher, president of the company, had been experimenting for some time with various chemicals with results so satisfactory to him that he decided to form a company for the purpose of producing rain by artificial means, making contracts for the same and doing business.

The third company to organize in Goodland was the Goodland Artificial Rain Company, chartered February 11, with J. H. Stewart as president. Its capital stock was $100,000 and its purpose as stated in the charter was "to furnish water for the public by artificial rainfall by scientific methods and to contract for services for the same, and to sell and dispose of the right to use our process in any city, township, county, state, territory or country." All these companies claimed to use the Melbourne method of producing rain.

As the word spread inquiries came pouring in from all directions for information on the Goodland rain companies and their operations. Soon a bidding war erupted between them driving prices down and igniting a frenzy for their services with contracts being signed left and right. One such contract was to produce half an inch of rain over Jewell County, Kansas, within five days for $500. An enterprising merchant took advantage of the anticipated arrival of the Rain Maker with the following advertisement in the local paper:

THE RAIN MAKER IS HERE
Call early at our Store and buy one of our Silk, Serge, or Satine Umbrellas. If you want to use them for sun umbrellas they especially answer that purpose.

As one would expect the press, then as now, tended to ridicule the proceedings. The *Jewell County Monitor* of 29 June declared that when the Rain Maker began "there were a few clouds in the sky, but he got his machine bottom-side up and dispelled the few there were and at present writing it is clear as a bell." *The Jewell County Republican*, of Jewell City, on 1 July was of the opinion that the Rain Maker was "simply betting his time against $500 that it will rain between Monday and the Sunday following."

"But," Ms. Caldwell tells us, "as the paper went to press it announced that a good rain was falling. Four days after

Mr. Murphy began operating a copious rain fell over Jewell County. His contract having been fulfilled he received four hundred dollars. A few failing to come up to their promise, he threw off one hundred dollars."

By the first week in July the three rain companies were at work in various locations across Kansas. One company in Jennings got rain, but a dispute arose as to the pay. The contract had called for a half inch in and around the town. While that amount did fall on either side, it fell short within the town, hence some townsmen refused to pay their pledges. Meanwhile in St. Francis, another company produced rain with it falling in quantities "never before seen in this county" at that time of the year. Doctor Swisher at yet another site failed to produce rain in the stated amount and thereby forfeited his pay. Swisher took full credit for all the rain that fell in nearby Thomas County, however. One editor remarked that "Hereafter Providence will get credit only for hail-storms and cyclones, but in time it is expected that the Goodland Rain Makers will take full charge of the universe."

With the coming of the rain, natural or artificial, and soon after it autumn, attention drifted to politics, 1892 being an election year. But an unusual dry spell in May 1893 revived the subject of rain making, and another player from Goodland appeared on the scene.

C. B. Jewell was the chief train dispatcher for the Rock Island railroad. He was stationed in Goodland and had watched the Rain Maker hysteria at close hand. He had been quietly conducting his own experiments since Melbourne's visit. He was convinced that he had discovered Melbourne's secret, and using his spare time had pursued his investigation with seeming success. In the spring of 1893 his experiments attracted the interest of officials with the Rock Island Road, and they offered to furnish him everything necessary to conduct his work and to make a thorough test of his theory. He was provided with a partitioned boxcar to be both his laboratory and living quarters. The company also furnished him with balloons for experimenting with the concussion theory. He was granted full freedom to experiment at points along the line.

Ms. Caldwell continues:

On April 30 the Rock Island sent Mr. Jewell $250 worth of chemicals, and on the following day, he with his assistant, Harry Hutchinson, began experimenting at the Goodland depot. The cool nights hindered the work somewhat, but on Wednesday a heavy rain fell in the southern part of the county, and on Friday a general rain began to fall, continuing in showers until Sunday noon. It was said to be the first general rain since August. Mr. Jewell, of course, claimed that the rain was the result of his efforts, and it was difficult to prove the contrary.

The Rain Makers now started out along the road making experiments at various places. They arrived at Meade Center on June 1. Here the Rock Island people made extensive preparations for the visit. Invitations, extended to the citizens of Dodge City, were accepted by Mayor Gluck, G. M. Hoover and many other persons. Instead of the air of mystery and secrecy maintained by other Rain Makers, Mr. Jewell allowed visitors in his laboratory and explained to them his methods, with the exception of revealing the materials used and the manner of compounding them. He explained that he used four jars to generate the gases, and utilized the circuit batteries to establish electrical communication with the clouds. On June 2 a light rain began falling, but not in sufficient quantities, and dynamite was fired into the air to assist the gases. Mr. Jewell wired the general superintendent of the road that the wind was blowing too hard to produce rain at Meade, but that a rain should fall in the vicinity of Salina. This happened as predicted, as on the next day a heavy rain was reported to have fallen there.

From Meade the Jewell company proceeded to

Dodge City, arriving on June 6. They began work at once surrounded by a crowd of spectators. People had driven for miles to witness the experiment. The natural condition of the atmosphere, being unfavorable for rain, gave the Rain Makers a chance to work on their own merits. On June 7 a representative of The Globe Republican visited their car and expressed his surprise that "they did not wear bald heads, long beards, nor forms bowed down by years of accumulated wisdom, but were a couple of hale and hearty young men," with frank, unreserved manner. He found "no air of mystery or complicated contrivances calculated to mystify the people." The experiment, nevertheless, was not a success. A high wind blowing continually from the time they commenced, scattered the gases and only a sprinkle fell on Thursday night. The Rain Makers claimed to be responsible for a rain that fell at Meade, and so were not disheartened at their failure at Dodge City, explaining that the wind carried the chemicals several miles and that the rain did not fall in any quantity where they were sent up. "Only on a calm day," said Jewell, "will it rain at the point where the experiment is made."

Mr. Jewell continued working locations all along the Rock Island Route throughout that summer. As could be expected his results were mixed. He also tried a variety of techniques, such as using rockets filled with his rain making chemicals.

The three rain companies of Goodland that had received so much publicity during 1892 were completely eclipsed by C. B. Jewell. A. B. Montgomery, an officer of the Inter-State Rain Company, continued to be enthusiastic though. In 1893 he attempted to secure a patent on the Inter-State Rain Company's rain making process. Ms. Caldwell tells us that:

At a convention in Wichita he explained that his company had operated in Sherman County but three times during the season. In July when the hot winds were about to ruin the wheat the company began operations and within twenty-four hours the hot winds had ceased and the temperature had dropped, and on the fourth day two inches of rain fell. The company made this experiment at their own expense, the people refusing to contribute. As a result Mr. Montgomery stated that Sherman County raised 100,000 bushels of wheat, and none was raised for a hundred miles east and west of them.

Despite Jewell's iffy results the Rock Island Road launched experiments on a larger scale in the spring of 1894. The railway outfitted three new rail cars and put them in the care of C. B. Jewell, Harry Hutchinson and W. W. LaRue. In April Mr. Jewell went to South Dakota in his new lab. There he gave instructions to several parties who had purchased his method. By chance or design, it rained every day he was there. The following messages were reported to have gone over the wires:

C. B. Jewell, Aberdeen. How much will it cost to stop this rain? Have a flock of calves in danger of drowning. V. N. W.

V. N. W., Britton. Machine wound up for ninety days. Same price for stopping as for starting. Teach the calves to swim. C. B. J.

The worried rancher was not the only person getting concerned. Opposition to rain making was growing like an approaching thunderstorm. Ms. Caldwell tells us:

The second week in May the Rock Island Rain Makers began work. They planned to make the first trials simultaneously at Selden, Phillipsburg and Norcatur. These were to be free, but after that they intended to make contracts and charge for their experiments.

On May 10 the three cars departed. But opposition to their operation began to be registered. The people in the eastern part of the state protested against their coming there as they had too much rain already. And the farmers of Sherman County held several meetings and resolved: "That the experiments of the Rain Makers had been detrimental to the crop prospects and instead of any rain being produced the gases sent up had produced heavy winds and cold weather, and that a committee be formed and wait upon these gentlemen and notify them to quit the business."

The people were also complaining because they thought that the dry weather of the spring was a visitation of Divine displeasure and that God had withheld the moisture from that section because of the "impudence of man in trying to take control of the elements out of His hands."

Mr. Jewell began operating at Wichita, June 9, where he threatened to turn Douglas Avenue into a canal by the next day. His threat was fulfilled in a measure, for a heavy rain falling that afternoon and night put all the rivers and creeks out of their banks. At the same time Jewell operators were making tests at Peabody and Wellington with like success. At Peabody rain began at one o'clock Saturday afternoon and was reported to have been the heaviest in three years. It was, however, a general rain extending all over the West. Some gave the Rain Makers credit for it, but the skeptic insisted that it was "the work of the Lord."

In July the interest in rain making suddenly died out, giving way to a rising enthusiasm over irrigation. Only slight mention is made of the activities of the Rain Makers after this.

While interest in rain making dissipated in Kansas in 1894 interest in the subject never really went away. Indeed, Kansas is a veritable hotbed of weather modification activities today.

For example, it is home to the Western Kansas Weather Modification Program (WKWMP). Their main office is at the Kearny County Airport in Lakin, Kansas. During the crop-growing season the WKWMP's objectives are to reduce crop and property damaging hail and to optimize usable precipitation. Their website somewhat defensively states that they employ "leading edge seeding techniques and technology" and that their methods "have been scientifically researched and are mainstream within the weather modification community." In 2006 they operated 24 hours a day, 7 days a week from 18 April through 16 September to provide this service to 11 counties in Western and Southwestern Kansas.

While today the Rain Makers are caricatures of ridicule, many men of good judgment believed in them, among whom were officials of the Rock Island Railroad, whose faith in Jewell's theory had been enough for them to spend a large sum of money testing it. The government of the United States had likewise manifested a belief in the possibility of producing rain on demand.

As well as the technical issues of the efficacy of the technology and the religious questions of whether using it displeased God or Gaia, legal problems cropped up from the beginning as well. One of the first court cases recorded was when Dr. Swisher of The Swisher Rain Company of Goodland sued a town for failure to pay his fee.

Another nettling problem was what to do if the Rain Maker made too much rain, a dramatic example of which occurred in San Diego, California, in 1916 as described by Tim Swartz in his article "Meteorological Madness: Is Weather Being Used As The Ultimate Weapon?" in which he wrote:

> ...City officials ...offered rainmaker Charles Hatfield $10,000 to end their local drought. Hatfield proceeded to set up a series of 24-foot-tall towers that were topped with boiling vats of a secret combination of chemicals. Near-by, farmers heard explosions and saw flames, as smoke filled the cloudless sky and

16

chemical smells permeated the air.

Soon the clouds opened up and it began to rain. Not only did the reservoirs fill but the rivers flooded, several dams burst, and dozens of people died. Although Hatfield fulfilled his promise, city officials blamed the deaths on Hatfield and ran him out of town, without paying him.

To this day the courts have been extremely reluctant to rule on tort issues related to weather modification. What is recognized today as legitimate and effective scientific modification of the weather began in the United States in the late 1940s with the development of cloud seeding technology. From its inception the debate over cloud seeding mirrored that over the 1890s Rain Maker's methods and legal liabilities, as we will see in the next chapter.

CHAPTER TWO

WEATHER MODIFICATION IN THE TWENTIETH CENTURY

Unlike the smoking cauldrons of unnamed gases and the deafening concussions from explosions used by the Rain Makers of the 19th century, cloud seeding became a well established if not universally accepted process for artificially modifying the weather in the following century. With the effectiveness of cloud seeding demonstrated, other technologies soon followed.

This chapter will focus on the evolution of this technology in the final half of the 20th century and how it went from attempts to steer hurricanes to its use as a weapon during the Vietnam War. Let's start with cloud seeding as it stands today, and then take a look back to the beginning of scientific modification of the weather…

CLOUD SEEDING

The idea is simple, and can be accomplished by several methods. The basic mechanism is to inject a substance into a cloud which forms a nucleus for ice to freeze around, creating something heavy enough to fall to earth as rain. Silver iodide is the best known and most commonly used substance for cloud seeding, but many others have been tried. Ice freezing nuclei have the effect of creating rain, reducing hail, and possibly preventing rain by over-seeding (as Otis Port reported the Russians did to keep it from raining on their May Day parade).

Let me introduce you to the privately owned firm of Weather Modification, Inc. (WMI) of 3802 20th Street North, Fargo, North Dakota. WMI says its Mission is:

> ...to provide solutions to the diverse water-management needs of its customers using superior airborne and ground-based technologies, accommodating innovative instrumentation and adverse conditions. WMI strives to provide the best cloud seeding services, training, and equipment in the world. All of this we do with the highest regard for safety, the environment, and the public well-being.

Their website lists several current projects including hail suppression in Argentina, snowpack augmentation in Idaho, and cloud seeding in Nevada. Their site also proudly states:

> Since Weather Modification, Inc. was founded in 1961, we have become the world leader in hail damage mitigation, precipitation augmentation, and the application of airborne remote and in situ sensing. This includes the instrumentation and expertise required for cloud microphysical and atmospheric aerosol sampling. We provide services to universities, governmental agencies, and private sector entities.

One of the programs WMI is a part of is The Oklahoma Weather Modification Program (OKWMP). It is an operational cloud-seeding program with the dual goals of rainfall enhancement and hail suppression. WMI, working as an independent contractor, has provided the personnel, aircraft, and associated equipment, as well as having conducted the seeding operations since the program's beginning as the Oklahoma Weather Modification Demonstration Program (OKWMDP) in 1996.

The purpose of the OKWMDP was to evaluate the potential effects and benefits of weather modification in Oklahoma. After three years of evaluation the state government decided to implement the program as an official element of Oklahoma's water management strategy. A state law was passed in 1999 creating a new Weather Modification Division within the Oklahoma Water Resources Board. The law also created a Weather Modification Advisory Board to oversee weather modification efforts.

The OKWMP's website (www.evac.ou.edu/okwmdp/) exists to provide some background information on cloud seeding, and to illustrate some of the results of the evaluation done by the Environmental Verification and Analysis Center and the Oklahoma Climatological Survey at the University of Oklahoma.

WMI is but one of many private firms, academic institutions and governmental agencies in the United States engaged in weather modification. Yet to this day many in mainstream science and the media (and "know-it-alls" in bars and barbershops) are unwilling to admit that this technology is any more than the smoking cauldrons of the 1890s Rain Makers.

Joe Gelt in his 1992 article for *Arroyo* magazine, "Weather Modification: A Water Resource Strategy to be Researched, Tested Before Tried," expressed this general unwillingness to accept weather modification saying:

> In some ways, weather modification or precipitation enhancement remains an idea whose time has not yet come. It has always had its advocates, but widespread acceptance has not followed for a variety of reasons. Even its supporters generally agree that weather modification has a public image problem. When explaining this cautious, wary attitude, analysts point to various scientific, socio-economic, and political factors.
>
> Some scientists and water resource managers are wary because they believe that weather modification research conducted thus far shows inconclusive

21

results. They seek additional scientific investigations to demonstrate quantitatively the benefits of cloud seeding. These people represent the jury that is still out.

Others are wary of weather modification because it goes against the grain of a certain ecological ethic. It represents an interference with a natural process, with results possibly difficult to predict and control. Man as geologic force built dams and controlled the course of powerful rivers, upsetting along the way ecological balances and causing environmental harm. What then might man as an atmospheric force accomplish?

The courts and regulatory bodies have had a similar problem coming to grips with the reality of weather modification, as mentioned in the previous chapter. In his article, Gelt tells us:

> Oft-lamented is the fact that everybody talks about the weather but no one does anything about it. To do something about the weather however is to raise various complex legal and public policy questions. For example: Who is liable for damages from floods or other weather events resulting from weather modification? How are the rights of those who want rain to be reconciled with the rights of those who prefer sunshine? What if precipitation increases in a basin in which cloud seeding occurred but decreased during the same period in another basin? Has the latter basin been wrongfully deprived of its rightful precipitation?
>
> And there are other questions: How is it determined that precipitation was in fact the result of weather modification? How is the amount of new water to be quantified for credit and distribution? On what basis is the new water induced by weather

modification to be allocated among water users? How can those who pay for the weather modification be ensured that they will in fact receive their share of the new water?

Such issues are the stuff and drama of lengthy and interesting court proceedings and water policy debates.

Also not to be neglected are possible environmental problems resulting from weather modification. Local or regional manipulation of climate could impact present plant and animal populations. For example, increased precipitation might mean increased weed growth, and a heavier snowpack could disrupt the winter food habitat of large mammals. Concern has also been expressed about the effects of introducing artificial condensation nuclei (e.g. silver iodide, dry ice and liquid propane) into the atmosphere.

Dr. Ronald B. Standler is an attorney in Massachusetts who earned a Ph.D. in physics in 1977. His first peer-reviewed scientific publication was a paper that reviewed the published literature on the toxicity of silver iodide used in cloud seeding. He did scientific research in atmospheric electricity and lightning before turning to the law. As an attorney in private practice he concentrates in computer law, higher education law, and copyright law; but he also does consulting with other attorneys, principally on scientific evidence in torts involving technology (e.g., damage by either lightning, electrical surges or "power quality," computer hardware and software, product liability, etc.). His article "Weather Modification Law in the USA" is a discussion and analysis of court cases in the US involving weather modification, and contains a detailed review of tort law in the US that applies to weather modification. In it he tells us:

> The history of cloud seeding also makes an interesting case study in the interaction between

23

scientists and society: not only about the obligations and ethics of scientists, but also about how courts have avoided deciding cases involving technical issues about weather modification.

It is clear that man already has the technology to modify weather and that more effective technology can be designed. However, we need scientific knowledge to understand how and when to use such weather modification technology, so that intelligent choices can be made, instead of guesswork. Civilization would immensely benefit if damage from drought, floods, hurricanes, hail, tornadoes, etc. could be reduced. But before we reap such practical benefits, we need much more basic scientific research.

Despite the potential immense economic importance of cloud seeding and the existence of commercial cloud seeding technology since 1950, the courts in the USA have not yet begun to resolve legal issues involving either negligent cloud seeding or the rights of landowners to rain from the clouds that are either above their land or upwind from their land.

Dr. Standler's companion essay "History and Problems in Weather Modification" not only gives the downside of developing this technology but also exposes the need to give adequate long-term financial support to basic scientific research *before* engaging in practical applications. He begins it with:

It is a common misconception that pure water freezes at a temperature of zero Celsius (32 degrees Fahrenheit). Zero Celsius is actually the temperature at which ice melts. Water freezes at a temperature between 0 and -39 Celsius, depending on the type of nuclei (i.e., contaminants) present. Liquid water with a temperature of less than 0 Celsius is called "supercooled water".

In November 1946, Dr. Bernard Vonnegut discovered that microscopic crystals of silver iodide (AgI) nucleate water vapor to form ice crystals. Vonnegut choose AgI crystals because there is nearly the same distance between molecules in the crystal lattice for both ice and AgI, which makes AgI the optimum material to nucleate ice.

Vonnegut not only discovered the ice-nucleating properties of AgI, but he also invented a practical way of generating tiny AgI particles to serve as nuclei for ice crystals. Vonnegut dissolved a mixture of AgI and another iodide in acetone, sprayed the solution through a nozzle to make droplets, then burned the droplets. More than fifty years later, Vonnegut's method continues to be the common way to seed clouds.

And concludes it with:

Basic scientific research should occur first. Only after the applicable scientific principles are understood can we have a rational application of law to weather modification, such as determining in tort litigation if a cloud seeder caused a flood or drought, or determining if a cloud seeder was negligent. Good laws and good regulations cannot be based on possibilities and conjectures.

Today there are a variety of cloud seeding technologies employed commercially around the world in attempts to:
- clear fog from airports,
- augment snowpacks in mountain regions,
- increase rain from summer showers,
- reduce destruction from hail, and to
- put out forest fires.

[Sidebar: If the name Vonnegut seems familiar that's because Dr. Bernard Vonnegut (29 August 1914 – 25 April

1997) was the younger brother of the famous author Kurt Vonnegut, Jr.]

Of course, cloud seeding using Dr. Vonnegut's discovery is not the only version of cloud seeding technology that has been tried.

A few of the most important patents in other cloud seeding technologies include the following:

U.S. Pat. No. 5,174,498 is for material useful for seeding supercooled clouds in order to augment rainfall. The material used is defined as an aliphatic long-chain alcohol (aliphatic means it is an organic chemical compound in which the carbon atoms are linked in open chains).

U.S. Pat. No. 4,600,147 is for a cloud seeding method of inserting liquid propane from a rocket to generate large numbers of ice crystals in supercooled clouds.

U.S. Pat. No. 5,357,865 includes the use of a pyrotechnic composition such as potassium chlorate or potassium perchlorate, which act as nuclei for precipitable water drop formation.

U.S. Pat. No. 4,096,005 involves another pyrotechnic cloud seeding composition comprising silver iodate and a fuel consisting of aluminum and magnesium.

One that has gotten quite a bit of media attention lately is United States Patent 6,315,213, granted 13 November 2001 to Peter Cordani, CEO of a company called Dyn-O-Mat. This patent is for a "method of modifying weather" by seeding rain clouds with a "suitable cross-linked aqueous polymer," a product that Dyn-O-Mat, a Florida-based environmental products company, calls Dyn-O-Gel. This polymer is dispersed into the cloud and the wind of the storm agitates the mixture causing the polymer to absorb the rain. This reaction forms a gelatinous substance that precipitates to the surface below, thus diminishing the clouds ability to rain. One beauty of this technique is that it does something that none of the others claim to be able to do—making all this "Jell-O" in the air should also reduce the velocities of the winds associated with the storm by up to 20 mph. And yes, there are a lot of potential downsides to this technology.

On 13 July 2001 Cordani and his team loaded 20,000 pounds of their product into a C-130 at Palm Beach International Airport and headed for a developing thunderstorm. [Sidebar: the heavy lift propeller driven Lockheed Martin C-130 Hercules is the U.S. Air Force's principal tactical cargo and personnel transport aircraft. It has been in continuous production since 1954 with over 2,260 having been built for 67 countries]. Soon thereafter the Dyn-O-Mat team made a bank of clouds simply disappear, a first-ever feat documented by Doppler radar. The plane scattered the Dyn-O-Gel powder through a storm cloud 1600 meters long and over 4000 meters deep. It took about 4000 kilograms of powder to soak up the moisture from the cloud, making it virtually disappear.

"I had calls from a weather tower and even from Channel 5 news in Miami, saying that they had seen the cloud literally disappear off the radar screen. They confirmed that there had been a tall build-up and the next moment it was gone," Cordani said.

According to the Associated Press, the American Meteorological Society says no hypothesis for hurricane modification has ever been proven to work. Meteorologists, AP reported, remain intrigued by Dyn-O-Mat's idea, but the field's history of unfulfilled promises has left them wary of Cordani's claims. "The Dyn-O-Mat folks need to develop a credible scientific hypothesis and move beyond anecdotal accounts [of] 'We dumped the stuff in a cloud, and it went away,'" says Hugh Willoughby, director of the Hurricane Research Division at the National Oceanic & Atmospheric Administration (NOAA).

Although the Dyn-O-Mat company claims their gel is 100% safe that seems hardly likely. The Dyn-O-Gel patent states that one of the ways it could be made safe would be to use a family of organically based polymers — but the actual chemical components of their gel, which remain undisclosed, are not organic. The patent also states that the "superabsorbent polymer" that Dyn-O-Gel is made of "is a resin capable of absorbing water up to several thousand times its own weight." So what happens when you swallow some? Would one swell up like the little girl who ate the "three course meal" bubblegum in the 2005 film *Charlie and The Chocolate Factory*?

Kidding aside, there are some real concerns. The United States Environmental Protection Agency (EPA), for example is concerned about potential fibrosis of the lung or other pulmonary effects that may be caused by inhalation of respirable particles of water-insoluble polymers. The toxicity may be a result of "overloading" the clearance mechanisms of the lung. EPA also has concerns for water-absorbing polymers based on data showing that cancer was observed in a two-year inhalation study in rats exposed to a water-absorbing polymer. Dyn-O-Mat insists that its product would not harm humans because it would be used only over the ocean. So what about dolphins and other aquatic mammals, or other species for that matter?

Nor has cloud seeding, by whatever means, been the only technology tried over the years since the days of the Rain Makers.

OTHER TECHNOLOGIES

Captain Howard T. Orville, USN is today immortalized by having a piece of the Antarctic landscape named after him. As the head of the Naval Aerological Service he was largely responsible for formulating the meteorological program for the Ronne Antarctic Research Expedition (RARE), led by Finn Ronne.

Ronne was born in Horten, Norway, on 20 December 1899. He studied at the Horten Technical College before immigrating to the United States. After working for the Westinghouse Electric Company, Ronne joined the United States Navy. He was a member of the Antarctic expeditions led by Richard Byrd in 1933-35 and again in 1939-40. Ronne then led his own expedition, RARE, to the South Pole in 1947-48. On his return he was able to announce that his research showed that

Antarctica was a single continent and not two islands. Finn Ronne died at Bethesda, Maryland, on 12 January 1980.

Ronne thanked Captain Orville for his invaluable assistance to RARE by naming a portion of the coast of Antarctica laying west of the Ronne Ice Shelf between Cape Adams and Cape Zumberge (at 75°45′ S 65°30′ W) the Orville Coast. Later Capt. Orville became the official White House advisor on weather modification.

President Eisenhower created the U.S. Advisory Committee on Weather Control in 1953. Captain Orville was appointed chairman of what soon became known as "The Orville Committee." Orville had previously been on the steering committee of Dr. Irving Langmuir's *Project Cirrus*, and his appointment was *de facto* confirmation of the success of Project Cirrus, which we will examine in detail shortly. In 1957 the Orville Committee's report explicitly recognized the military potential of weather modification, warning that it could become a more important weapon than the atom bomb!

Capt. Orville is widely quoted as having described for the press in 1958 a study by the Department of Defense, in which he said the DoD was studying "ways to manipulate the charges of the earth and sky and so affect the weather" by using electronic modification of the atmosphere to ionize or de-ionize selected regions of the sky over targeted areas. It would seem that nearly 50 years later the Russians put the DoD's idea to work—in Mexico!

The *Wall Street Journal* on 2 October 1992 reported that a Russian company called "Elate Intelligent Technologies, Inc." sells weather control equipment by using the advertising slogan "Weather Made to Order." The commercial Director of Elate, Igor Pirogoff, stated that "Elate is capable of fine-tuning the weather patterns over a 200 square mile area for as little as $200 U.S. per day." In Otis Port's "Rainmaking Has Its True Believers—And Skeptics" we read that:

Perhaps the most controversial technology comes

29

from Russia and Mexico. In 1996, Russian space and weather-control scientists hooked up with Gianfranco Bisiacchi, then head of Mexico's space efforts, and founded Electrificación Local de la Atmosfera Terrestre (ELAT). Nominal results from the three ground stations set up by ELAT in 1998 were so impressive — rainfall was reported to increase by as much as 30% — that Mexican state governments were soon clamoring for more facilities. There are now 13, with additional ones being installed in Baja California and the state of Puebla.

ELAT claims credit for ending the severe drought in northern Mexico. Since 2000, says Bisiacchi, the amount of annual rain in the region has been "30% to 35% greater than what it was during the 1990s. In fact, the lakes of the region that were dry are now full." When operations in the northern states of Sonora and Chihuahua started in 2004, he adds, most lakes were around 8% full. "We've now gone to levels of 85% to 90% — in just one year."

ELAT says its technology is more efficient than regular cloud-seeding methods. "Milking" clouds is usually done by sprinkling them with particles of silver iodide. The particles provide a site where the clouds' ice crystals accumulate in clumps too heavy to stay aloft. Bisiacchi and his team take a different tack: They generate charged ions on the ground and point them skyward. That, they claim, fosters clumping on both airborne dust particles and ice crystals touched by a charged ion.

This technology has plenty of skeptics. "Personally, I think it's a hoax," says Roelof T. Bruintjes, a weather-modification expert at the National Center for Atmospheric Research (NCAR) in Boulder, Colo. "It has no scientific basis."

Bisiacchi isn't fazed. "Look," he says, "this is new technology — a new scientific hypothesis. It's the same whenever you try to do something really new."

The *Wall Street Journal* article quoted Elate Director Pirogoff as saying that Hurricane Andrew could have been decreased "into a wimpy little squall"!

Many Americans, like my father for one, blamed the weird weather of the mid-20th century on atmospheric nuclear testing; but few knew about the many other attempts at manipulation and control of the atmosphere. The government and other interested parties have continued to do strange things in, and to the sky since.

Starting in the 1960s a number of significant upper-atmospheric experiments took place. In one experiment copper needles were dumped into the ionosphere as a "telecommunications shield." The government claimed they were trying to create an artificial ionosphere so as to maintain radio communications without interruption. The natural ionosphere is regularly affected by a number of factors, particularly sunspot activity, which can adversely affect radio reception. They attempted to put 3.5 billion copper needles, each 2 to 4 cm (about an inch) long, into orbit at an altitude of 3,000 km up. These were supposed to form an artificial uniform ionosphere about 6 miles thick and 25 miles wide. The U.S. planned to add to the number of needles if the experiment was successful. This plan was strongly opposed by the International Union of Astronomers.

Dr. Nick Begich and Jean Manning, in their book about the High-frequency Active Auroral Research Program (HAARP), *Angels Don't Play This HAARP*, shared a quote from private correspondence that they had had with Leigh Richmond Donahue, a researcher with the Centric Foundation of Maggie Valley, North Carolina. In writing about Ms. Donahue they said:

> [She] tracked events during the postwar years and through 1977 alongside a physics genius, her late husband Walter Richmond. She writes, "...when the

military sent up a band of tiny copper wires into the ionosphere to orbit the planet so as to 'reflect radio waves and make reception clearer' we had the 8.5 Alaska earthquake, and Chile lost a good deal of its coast. That band of copper wires interfered with the planetary magnetic field."

Another series of experiments that should have produced environmental concern were the barium releases, which may still be ongoing. For example, on 12 December 1980, a rocket with a 48 kg barium payload was launched from the A-15 site of the Eglin/Santa Rosa Island Test Range in Florida. The barium was released in the F-1 region of the ionosphere at an altitude of 182.7 km. Approximately 32 minutes later a second rocket was launched, carrying a diagnostics complement which included a pair of pulsed-plasma-probes and an ion mass spectrometer for direct measurements of electron density, temperature, ion composition and other factors. This was just one of literally hundreds of intentional barium releases by rockets and balloons over a period of thirty years or more. These were alleged to be in aid of better understanding the dynamics of upper atmospheric physics.

Another such release was a barium cloud test in Peru in 1979 conducted by folks from Air Force Wright Aeronautical Labs of Wright-Patterson Air Force Base in Ohio. In an effort to investigate the mechanism causing ionospheric irregularities to form in the equatorial region an experiment was designed by the Max Planck Institute involving the release of two barium clouds in the F region of the ionosphere just after sunset. The electric field resulting from the two barium clouds was expected to trigger an ionospheric irregularity that would propagate up through the ionosphere as a depletion bubble. The barium was accidentally released 50 km below the planned altitude so no significant ionospheric irregularities were formed. The Max Planck Institute of Germany is one of the world's leading physical sciences research institutions. They have operated a HAARP-like ionospheric heater in

Norway since the 1970s.

The Naval Research Lab (now the Office of Naval Research) of Washington, D.C. and numerous subcontractors experimented for decades with using these barium releases to create clouds that could be used to enhance radio communications. Mission Research Corporation of Santa Barbara, California conducted one such trial. In a series of experiments conducted between April and July 1978 in conjunction with an ongoing series of barium releases an experiment was undertaken to determine if radio communication was possible off the base of striated plasma created by these releases. A transmitting station was set up to broadcast a steady signal at two different HF frequencies (high frequency, the same range HAARP broadcasts in) toward the base of the barium striations and two receiving stations listened for signal returns on the two frequencies. One station heard substantial returns while the other heard nothing. HAARP, as we shall see was an outgrowth of studies like these.

Unfortunately barium is toxic to humans. In an article for the November 2000 issue of the *Idaho Observer* Amy Worthington wrote:

> Leading chemtrail researcher Clifford Carnicom has completed a series of impressive reports citing evidence that our atmosphere is now saturated with barium compounds as a result of the military's on-going weather and atmospheric modification projects. The presence of metallic alkaline salts in rainfall samples collected nationwide indicates that the atmospheric pH is being rapidly modified — most likely by barium.
>
> We know that America's military-industrial complex has been spewing various forms of barium into our atmosphere for years. The University of Alaska has propelled barium into space in order to study the earth's magnetic field lines. ...A recent

report from Wright-Patterson Air Force Base confirms that the Air Force has been spraying barium titanate across the United States to facilitate advanced radar studies.

Chemical handbooks state that barium is highly toxic to human beings. The officially "safe" levels of barium in the environment are quite low, on the order of 1-2 parts per million. The Agency for Toxic Substances and Disease Registry warns that humans who ingest high levels of barium can develop problems with the heart, stomach, liver, kidneys, spleen and other organs. It also confirms that ingesting high levels of water soluble barium compounds can cause:

- difficulties in breathing
- changes in heart rhythm
- increased blood pressure
- stomach irritation
- brain swelling
- muscle weakness

Soluble salts of barium can stimulate all muscles of the body, producing contractions of the skeletal muscles and spasms of the smooth muscles of blood vessels, bronchi, stomach and intestines. These salts can radically increase the force of the heartbeat, a potentially lethal situation for the elderly and the chronically ill. In toxic doses, these salts can cause high blood pressure, asthmatic attacks, burning sensation in the stomach, nausea, vomiting and convulsions. One chemical directory advises that barium be kept out of the reach of children. Great stuff to be spraying over the civilian population of our nation, is it not?

Many other technologies, often with even less recognition from mainstream science than the Rain Makers of old received, have been used in attempts to manage or control various aspects of the environment and the atmosphere, especially in rain making. One such off-trail "science" is called *Radionics*. In his article "Radionics: At the Crossroads of Science & Magic" Thomas Brown tells us that:

> Radionics is a methodology for the detection and manipulation of subtle energies linked to physical matter: animal, vegetable or mineral. It is generally known as a system of vibrational healing wherein curiously configured "black boxes" are used to detect and treat disease conditions of a living body, be it animal or vegetable.
>
> Radionics instruments can also broadcast healing vibrations to a single subject. They can be used to detect specific elements in mineral samples, or even to test for mineral veins using an aerial photograph or map. Their potential makes them appear magical when considered from the viewpoint of modern science.
>
> Radionics instruments consist generally of a circuit of resistors or capacitors arranged in a specific pattern or order, which provides a stable system of "rates" which may be dialed in to set the instrument for various organs or disease patterns. The mystery of the black boxes is no mystery once one understands the relationship between form, matter and energy. They are specialized tuning circuits, operating on energies outside of the electromagnetic spectrum. Their electronic structure makes no sense to the engineer. The history of modern Radionics spans the 20th century back into the 19th, and many serious medical workers have used it as a viable system of diagnosis and treatment. It has come under fire from medical authorities in many countries, and at times with good reason. Its workings are beyond the understandings allowable in the modern scientific worldview. There are many individuals and groups

around the world who quietly use this technology successfully, regardless of whether or not it can be explained to the rational mind.

...The origin of Radionics is found in the ancient art of radiesthesia, or dowsing. Dowsing is a form of detection of energies and force fields through reactions to subtle vibrations. Radionics instruments are designed to aid the investigator in detecting and identifying subtle vibrations. In dowsing the pendulum or divining rod acts as an amplifier of the body's minute reactions to the sought after goal, be it water, oil or a buried pipe. The dowsers hold in their minds the sought after goal and their body reacts and the pendulum amplifies the reactions.

Dr. Andrew Michrowski, Ph.D. and Peter Webb gave a presentation, with the title "Radionics Case Study — Rain for Gujarat State, India," to an audience at the 2005 conference of the United States Psychotronics Association. Dr. Michrowski is the director of Planetary Association for Clean Energy (PACE) and a frequent presenter at Radionic and Psychotronic conferences. Peter Webb is the President of Laurentia Inc., a designer and manufacturer of Radionic devices. In this talk they describe how one of Peter Webb's machines brought rain to a drought ravaged region of India.

In 2004 the monsoon rains came but missed the state of Gujarat in Northern India and a full blown drought ensued with crops failing and so on. With a low water table from over-pumping the aquifer it was a disaster in the making. Throughout that monsoon season all stormy weather had stalled south of some sort of atmospheric pressure barrier around the middle of India.

At the time that Peter Webb and his team were called upon to relieve the drought the only weather system with significant moisture appeared to be hovering over The Sudan and Eritrea in Eastern Africa, thousands of kilometers to the west of Gujarat. To attract this system, as well as to assist the northerly movement of moisture that lay southwest of the Indian subcontinent in the Indian Ocean, Peter Webb, using his many years of expertise and his trusty Radionics unit, created a low-pressure system above Gujarat. After five days the desired system was pulled from Africa and arrived over Gujarat, subsequently pouring torrential rain on this beleaguered area — not only relieving the drought but frightening the locals to the point where they begged to have the Radionic transmitter turned off! Where have we heard *that* before?

As I am sure you could have guessed, mainstream science is convinced that radionic manipulation of the weather as described above is completely impossible. The official position of the World Meteorological Organization, as restated by Michael J. Manton, Chief of Division of the Australian Bureau Of Meteorology Research Centre is:

> "The energy involved in weather systems is so large that it is impossible to create artificially rainstorms or to alter wind patterns to bring water vapour into a region. The most realistic approach to modifying weather is to take advantage of microphysical sensitivities wherein a relatively small human-induced disturbance in the system can substantially alter the natural evolution of atmospheric processes."

Which is to say cloud seeding is the only technology that is even grudgingly recognized by meteorologists. Many other technologies, some highly questionable, some established if not universally recognized could be cited — but for the sake of brevity let us return to the central core of mainstream scientific environmental modification for the rest of this chapter. Feel free to research this further if you are interested in off-trail technologies — the work of Wilhelm Reich, James Trevor Constable, and James DeMeo would be interesting places to start.

As early as the late 1940s Dr. Wilhelm Reich, a psychoanalyst who escaped Nazi Germany to become America's most radical naturalist, was developing weather modification techniques at his Orgonon Research Center in Rangeley, Maine. The U.S. Food and Drug Administration imprisoned Reich in 1954 for a minor interstate transportation infraction committed by an employee and Reich subsequently died in federal prison in 1957. Charles R. Kelley's article "What Is Orgone Energy?" defines what Dr. Reich thought he had discovered, saying:

> Franz Anton Mesmer called it animal magnetism; Charles von Reichenbach called it odyle. To Henri Bergson it was the elan vital, the "vital force;" while to Hans Driesch it was the entelechy. Sigmund Freud observed its functioning in human emotions and termed it libido. William MacDougall, the great British-American psychologist of a generation ago, labeled it hormic energy. Dozens, if not hundreds, of lesser-known scientists have recognized its presence and have given it a name to characterize its special properties. Among the 20th century proponents of the concept are, for example, Doctors Charles Littlefield and his vital magnetism and George Starr White and his cosmo-electric energy . Mechanistic science in the 17th through 19th centuries embraced many of its essential qualities in the concept of the ether, while mystical human beings have embraced other essential qualities of it in the concept of god. Orgone energy is Wilhelm Reich's name for the substratum from which all nature is created. The best definition this author can provide for it is this: Orgone energy is the creative force in nature.

In 1954 all books dealing with *orgone* were banned and burnt by the U.S. government, and Reich was jailed and, many believe, assassinated. James Trevor Constable and James DeMeo are but two of the many authors and rainmakers who have followed in Reich's footsteps, possibly demonstrating the validity of his work.

HURRICANES: EYE OF THE WEATHER CONTROVERSY

Finding a way to steer or reduce the force of hurricanes has been a goal of some weather scientists, government agencies and commercial investors and insurers since the end of World War II. Those who would control hurricanes began with cloud seeding and have progressed to other developing technologies as they became available.

Dr. Ross N. Hoffman is a principal scientist and Vice President For Research And Development at Atmospheric and Environmental Research (AER) whose corporate headquarters is located at 131 Hartwell Avenue, Lexington, Massachusetts. Their hardly modest website describes the company as follows:

> Established in 1977, Atmospheric and Environmental Research, Inc. (AER) is a world-renowned research and development facility for cutting edge technology. AER's expertise ranges from weather and climate prediction and dynamics, to ocean modeling and data analysis, air quality and risk assessment, to remote sensing, sensor design and data analysis, modeling of planetary atmospheres, and evaluation of the effects of the atmosphere on communications systems. Clients include civilian and military agencies of the United States of America, as well as interests in the commercial, financial, energy, and insurance industries. With the top minds in research AER's applications have fundamentally impacted the way businesses, researchers, and national governments interpret the Earth around us.

Dr. Hoffman has been a member of several NASA science teams and was a member of the National Research Council

Committee on the Status and Future Directions in U.S. Weather Modification Research and Operations. His article "Controlling Hurricanes — Can Hurricanes And Other Severe Tropical Storms Be Moderated Or Deflected?" appeared in the October 2004 issue of *Scientific American.* It began with:

> Every year huge rotating storms packing winds greater than 74 miles per hour sweep across tropical seas and onto shorelines — often devastating large swaths of territory. When these roiling tempests — called hurricanes in the Atlantic and the eastern Pacific oceans, typhoons in the western Pacific and cyclones in the Indian Ocean — strike heavily populated areas, they can kill thousands and cause billions of dollars of property damage. And nothing, absolutely nothing, stands in their way.
>
> But must these fearful forces of nature be forever beyond our control? My research colleagues and I think not. Our team is investigating how we might learn to nudge hurricanes onto more benign paths or otherwise defuse them. Although this bold goal probably lies decades in the future, we think our results show that it is not too early to study the possibilities.
>
> To even consider controlling hurricanes, researchers will need to be able to predict a storm's course extremely accurately, to identify the physical changes (such as alterations in air temperature) that would influence its behavior, and to find ways to effect those changes. This work is in its infancy, but successful computer simulations of hurricanes carried out during the past few years suggest that modification could one day be feasible. What is more, it turns out the very thing that makes forecasting any weather difficult — the atmosphere's extreme sensitivity to small stimuli — may well be the key to achieving the control we seek. Our first attempt at influencing the course of a simulated hurricane by making minor changes to the

40

storm's initial state, for example, proved remarkably successful, and the subsequent results have continued to look favorable, too.

To see why hurricanes and other severe tropical storms may be susceptible to human intervention, one must understand their nature and origins. Hurricanes grow as clusters of thunderstorms over the tropical oceans. Low-latitude seas continuously provide heat and moisture to the atmosphere, producing warm, humid air above the sea surface. When this air rises, the water vapor in it condenses to form clouds and precipitation. Condensation releases heat — the solar heat it took to evaporate the water at the ocean surface. This so-called latent heat of condensation makes the air more buoyant, causing it to ascend still higher in a self-reinforcing feedback process. Eventually, the tropical depression begins to organize and strengthen, forming the familiar eye — the calm central hub around which a hurricane spins. On reaching land, the hurricane's sustaining source of warm water is cut off, which leads to the storm's rapid weakening.

Because a hurricane draws much of its energy from heat released when water vapor over the ocean condenses into clouds and rain, the first researchers to dream of taming these unruly giants focused on trying to alter the condensation process using cloud-seeding techniques — then the only practical way to try to affect weather.

Let us take a look at the disaster that was the first known American attempt to control a hurricane, *Project Cirrus*.

PROJECT CIRRUS

Project Cirrus was a joint operation conducted by the U.S. Navy and a group of scientists from General Electric (GE), led by Dr. Irving Langmuir. It was but one of many weather modification efforts in the U.S. to go horribly wrong—and in it we see that the U.S. government has had a policy of concealing the facts from the public since the inception of the use of this technology.

Appropriately enough the ill-fated Project Cirrus took place on the 13th of October in 1947 when a U.S. Navy plane, under the direction of the scientists from GE, flew into a hurricane and dropped 80 kg (176 lbs) of dry ice into it. At the time the hurricane was in the Atlantic Ocean, safely off the eastern coast of the U.S. After being seeded with dry ice the hurricane abruptly changed direction and came on shore near Savanna, Georgia, where it did extensive damage to property leaving 1400 people homeless and causing two fatalities. In what has since become almost standard procedure, the U.S. military classified the data from the seeding of this hurricane to frustrate litigation.

Dr. Langmuir believed that there was a 99% probability that the hurricane's change of direction was in fact the result of the cloud seeding! Langmuir's opinion about the effectiveness of cloud seeding, however, was never mentioned in any of his writings in scientific journals. It is in the classified 1953 final report on Project Cirrus, however.

There is also a 99% probability that attorneys for General Electric, seeking to avoid litigation from victims of the hurricane, reviewed and censored Langmuir's scientific publications and ordered Dr. Langmuir not to make any public admission that cloud seeding caused the hurricane to change direction. A biography of Langmuir says, "For the first time in Langmuir's long career at GE, officials occasionally wanted to know in advance what he was going to say in his public reports."

Keith Harmon Snow graduated B.S.E.E. and M.S.E.E. with a specialty in microwaves and antennas engineering from the University of Massachusetts, Department of Electrical and Computer Engineering, in 1986. From 1985 to 1989 he worked for General Electric Aerospace Electronics Laboratory on aerospace and defense technologies for

42

classified communications, RADAR, EW (Electronic Warfare) and Strategic Defense Initiative (SDI) programs.

Since 1990 Keith Harmon Snow has worked as a journalist. His articles and reports have appeared in publications in the US, UK and Japan. In Tokyo, he worked on an assignment for Newsweek, and he was staff writer, photographer & editor at Japan International Journal. He has three publications in the aerospace and defense journals of the Institute of Electronics and Electrical Engineers (IEEE).

Returning from his investigations of war in central Africa and the genocide in Rwanda, in 2001 Keith gave expert testimony, on genocide and U.S. covert operations in Africa, at a special congressional hearing in Washington D.C. He also attended the International Criminal Tribunal on Rwanda in Arusha, Tanzania. In 2002, two of Keith's reports on Central Africa won awards from the internationally recognized Project Censored and they are included in Project Censored 2003, a book on the top 25 underreported news stories of 2001-2002.

His website (allthingspass.com) has a 100-plus page report entitled: "Out of the Blue: Black Programs, Space Drones & The Unveiling of U.S. Military Offensives in Weather as a Weapon." In it he gives us an in depth analysis of military modification of the environment (ENMOD) from its origins to the present day. In it he wrote:

> GE's Dr. Irving Langmuir was soon testing a commercial cloud seeder in Honduras, in cooperation with the highly repressive United Fruit Company. United Fruit was a Rockefeller enterprise with close ties to the CIA, and Honduras was one of its Banana Republics. Setting the precedent for coming military efforts to downplay ENMOD successes, hide promising results, and deny information about ENMOD programs to the newly manufactured Red Menace of the cold warrior imagination, the U.S. government, with the help of GE lawyers, downplayed Dr. Langmuir's findings. His Project Cirrus report was

initially classified, to his consternation, and when the report was finally released to the scientific community, it contained a highly skeptical assessment by a panel of "experts," hand-picked by the DOD, who suggested that Langmuir's experiments were inconclusive, at best, that his science did not meet acceptable standards.

While publicly downplaying any conclusive evidence of ENMOD capabilities, however, the U.S. government embarked on a massive program of research and development. The liberal government funding, coupled with the explosion of ideas and research proposals, led to competing government agencies and overlapping programs.

The U.S. Navy and Air Force conducted numerous and systematic cloud seeding experiments from 1948 to 1950 and these demonstrated the initial promise of the ENMOD arena. From 1951-1953 they conducted the Artificial Cloud Nucleation Project (AEN), a large-scale project in southwestern Washington. By the early 1950s, some 10% of the entire land area of the U.S. was under commercial cloud-seeding operations, with some $3-5 million being expended annually. Public utilities like Pacific Gas and Electric (PG&E) and the Southern California Edison Co. maintained extensive programs throughout the 1950s and 1960s.

As early as 1953, Herbert Appleman of the U.S. Air Weather Service noted that contrails—condensation trails—formed by water vapor and other gaseous exhausts exiting the tailpipe of a jet engine can lead to cloud formation which may persist and spread out. Silver iodide generators mounted on airplanes took cloud seeding to new heights; unique but similar seeding results were attained after researchers at the Naval Research Laboratory (NRL) used carbon (black) dust dispersed by planes.

Commercial crop-duster pilots had previously been hired to seed with carbon black dust for the Air

Force—in the 1950s. The Air Force has not failed to notice that the 'solar absorption potential" of carbon black dust makes it an expeditious choice "to enhance rainfall on the mesoscale [atmospheric], generate cirrus clouds, and enhance cumulonimbus (thunderstorm) clouds in otherwise dry areas."

In 1953 and 1954, scientists at New York University seeded nineteen cyclones. In 1958 the Naval Research Laboratory (NRL) was conducting cloud dissipation experiments in Georgia. Cloud seeding became a routine operation for United Air Lines (UAL) planes in the 1960s. In 1959, the U.S. Congress established the Interdepartmental Committee for Atmospheric Sciences, a centralized command for ENMOD programs.

Under Project Skyfire in 1960 and 1961 the US Army pursued lighting suppression through experiments where millions of tiny metallic needles were released to 'seed" clouds. (These bits of foil are actually tiny dipoles whose ends are oppositely charged.) The dispersal of electronic chaff is today a major element in airborne missile evasion countermeasures, where the electronic tracking devices on an inbound missile are "fooled" by the chaff "decoy" materials launched by the pilot under attack. The U.S. Army Atmospheric Sciences Laboratory at Fort Monmouth, N.J., explored the potential of using chaff to pull the teeth of a thunderstorm. Project Skyfire scientists launched rockets to "trigger" lightning discharges. Project WhiteTop seeded clouds in Missouri throughout the 1960s.

Greg Machos at hurricaneville.com wrote at length about the history of massaging hurricanes. Picking up after Project Cirrus he wrote:

Nevertheless, Langmuir's work had generated

some enthusiasm. This enthusiasm was particularly strong among officials in the United States Government. In the years following Langmuir's experiment, a number of powerful hurricanes made landfall in the United States.

They included Carol, Edna, and Hazel in 1954, and the first billion dollar hurricane, Hurricane Diane in 1955. All four of these storms were Category Four strength according to the Saffir-Simpson Scale, and caused extensive damage from Florida all the way to New England and even Canada.

In response to these devastating storms, President Eisenhower appointed a Presidential Commission to investigate the idea of storm modification. Despite the lack of enthusiasm for the idea, Congress extended the life of this special committee for another two years in 1956, and by the end of the decade there were scientists that were ready to take another stab at attacking the hurricane.

In the early 1960s, two significant developments in the area of storm modification energized the quest to weaken and eradicate the hurricane. The first was the development of a new cloud seeding process by Dr. Robert H. Simpson, who was the director of the National Hurricane Research Labs in Miami, Florida.

Simpson, who eventually went on to develop the classification system known as the Saffir-Simpson Scale, theorized that hurricanes could be weakened by releasing frozen nuclei or particles of Silver Iodide compound (AgI) into the wall clouds of a hurricane or tropical storm, and imbalance the forces within the storm system.

Simultaneously, a group at the Navy Weapons Center in California improved seeding technology by developing new seeding generators that would be able to release large amounts of crystals into tropical storms and hurricanes.

As a result, Project Stormfury was born in 1962. Prior to that, a test case had already been done on Hurricane Esther in September, 1961, and with some success. The project team of workers from both the U.S. Weather Bureau and the Navy was able to decrease the sustained winds in the storm by ten percent.

PROJECT STORMFURY

One of the many current research facilities of the US government is the Atlantic Oceanographic and Meteorological Laboratory (AOML). It is one of the Oceanic and Atmospheric Research (OAR) Facilities of the National Oceanic and Atmospheric Administration (NOAA). NOAA/AOML is a part of the US Department of Commerce (DoC) and is located in Miami, Florida. AOML's mission is to conduct basic and applied research in oceanography, tropical meteorology, atmospheric and oceanic chemistry, and acoustics. The research seeks to understand the physical characteristics and processes of the ocean and the atmosphere, both separately and as a coupled system. AOML's website says that *Stormfury* was:

> ...an ambitious experimental program of research on hurricane modification carried out between 1962 and 1983. The proposed modification technique involved artificial stimulation of convection outside the eye wall through seeding with silver iodide. The invigorated convection, it was argued, would compete with the original eye wall, lead to reformation of the eye wall at larger radius, and thus, through partial conservation of angular momentum, produce a decrease in the strongest winds.
>
> Since a hurricane's destructive potential increases rapidly as its strongest winds become stronger, a reduction as small as 10% would have been worthwhile. Modification was attempted in four hurricanes on

eight different days. On four of these days, the winds decreased by between 10 and 30%. The lack of response on the other days was interpreted to be the result of faulty execution of the seeding or of poorly selected subjects. These promising results came into question in the mid 1980s because observations in unmodified hurricanes indicated:

• that cloud seeding had little prospect of success because hurricanes contained too much natural ice and too little supercooled water, and

• that the positive results inferred from the seeding experiments in the 1960s stemmed from inability to discriminate between the expected results of human intervention and the natural behavior of hurricanes.

Greg Machos continues:

After the creation of Project Stormfury, the new team assembled put together a reformulated idea on how to attack a hurricane. They proposed that a hurricane could be weakened by converting the supercooled water within the deep clouds of the storm is converted to ice, the hurricane's vertical column of air would be warmed and the storm would be weakened.

The team didn't have many chances to work on this new idea over the next eight years because there wasn't any storms that were far enough from land, and the team didn't want to risk litigation in case something went wrong. The team did have several chances though.

First, in 1963, they were able to conduct tests on Hurricane Beulah, but with only marginal success. Then, in 1965, the team considered seeding Hurricane Betsy, but due to the close proximity to Puerto Rico and other Caribbean islands, and the storm's erratic motion, the team did not go through with it, and Betsy ended up slamming into South Florida and causing severe damage.

Betsy was the last major hurricane to make a direct hit on South Florida before 1992 when Hurricane Andrew devastated Homestead, Florida and ended up causing $27 billion dollars in damage. Finally, in 1969, Project Stormfury was going to have a significant test case.

It was on the heels of Hurricane Camille barreling into the Gulf Coast regions of Mississippi and Alabama when Hurricane Debby was seeded on a couple of occasions over the two day period of August 19-20, 1969. Each time the storm was seeded, sustained winds were reduced significantly.

The first time, winds dropped 31 percent while the second time, they only dropped 15 percent. The apparent success with Debby helped fuel new projects, and improvements in technology. In particular, Hurricane Hunter aircraft, which went up dramatically during the 1970s.

Ultimately though, Project Stormfury was cancelled in 1980 since the team was unable to clearly ascertain whether or not the seeding efforts were really causing storms to weaken, or the systems just became victims of the environment around them. Nevertheless, the work done did bear some fruit as forecasters and scientists alike were able to learn a great deal from their research, and it has helped them improve forecasting accuracy.

As indicated above, scientific research into weather modification really took off in the 1970s. In part this was due to increasingly successful use of cloud seeding technology. The evolution of a scientific infrastructure was perhaps of even grater importance. Ever more powerful computers and sophisticated statistical approaches became available. Also, increasing scientific manpower kept pace with growing research budgets. This allowed for investigation and research into atmospheric dynamics of sufficient scope to pay off in producing meaningful scientific results. The weather balloons and sounding rockets of the fifties gave way to the geosynchronous satellites and ionospheric heaters of the seventies.

While the United States government has currently backed off from overt funding of this research, other nations and private corporations have picked up the slack. If anything, the pace of discovery in this field has only continued to accelerate into the 21st century. To continue to quote from Dr. Hoffman's article in *Scientific American*:

> A chaotic system is one that appears to behave randomly but is, in fact, governed by rules. It is also highly sensitive to initial conditions, so that seemingly insignificant, arbitrary inputs can have profound effects that lead quickly to unpredictable consequences. In the case of hurricanes, small changes in such features as the ocean's temperature, the location of the large-scale wind currents (which drive the storms' movements), or even the shape of the rain clouds spinning around the eye can strongly influence a hurricane's potential path and power.
>
> The atmosphere's great sensitivity to tiny influences—and the rapid compounding of small errors in weather-forecasting models—is what makes long-range forecasting (more than five days in advance) so difficult. But this sensitivity also made me wonder whether slight, purposely applied inputs to a hurricane might generate powerful effects that could influence the storms, whether by steering them away from population centers or by reducing their wind speeds.

Note Dr. Hoffman's repeated comments on how sensitive the atmosphere is to "tiny influences." Remember Dr. Gordon

J. F. MacDonald's statement that: "The key to geophysical warfare is the identification of environmental instabilities to which the addition of a small amount of energy would release vastly greater amounts of energy." Clearly if one can steer a hurricane away from a population center one can, using the exact same technology, steer it into one.

Dr. Hoffman was not able to pursue his ideas until computerized simulations became available in the last decade of the Century. With funding support from the NASA Institute for Advanced Concepts, he and his co-workers at the privately owned Atmospheric and Environmental Research (AER) have since employed detailed computer models of hurricanes in an effort to identify the kinds of actions that pay off in the real world. In particular, they used weather-forecasting technology to simulate the behavior of past hurricanes and then to test the effects of various interventions by observing the resultant changes in the modeled storms. Of his research Hoffman wrote:

> We have modified for our experiments a highly effective forecast initialization system called four-dimensional variational data assimilation (4DVAR). The fourth dimension to which the name refers is time. Researchers at the European Center for Medium-Range Weather Forecasts, one of the world's premier meteorological centers, use this sophisticated technique to predict the weather every day.
>
> After simulating a hurricane that occurred in the past, we can then change one or more of its characteristics at any given time and examine the effects of these perturbations. It turns out that most such alterations simply die out. Only interventions with special characteristics — a particular pattern or structure that induces self-reinforcement — will develop sufficiently to have a major effect on a storm. ...To explore whether the sensitivity of the atmospheric system could be exploited to modify atmospheric phenomena

as powerful as hurricanes, our research group at AER conducted computer simulation experiments for two hurricanes that occurred in 1992. When Hurricane Iniki passed over the Hawaiian island of Kauai in September of that year, several people died, property damage was enormous and entire forests were leveled. Hurricane Andrew, which struck Florida just south of Miami the month before, left the region devastated.

Surprisingly, given the imperfections of existing forecasting technologies, our first simulation experiment was an immediate success. To alter the path of Iniki, we first chose where we wanted the storm to end up after six hours--about 60 miles west of the expected track. Then we used this target to create artificial observations and fed these into 4DVAR. We set the computer to calculate the smallest change to the initial set of the hurricane's key defining properties that would yield a track leading to the target location.

The most significant modifications proved to be in the starting temperatures and winds. Typical temperature adjustments across the grid were mere tenths of a degree, but the most notable change--an increase of nearly two degrees Celsius--occurred in the lowest model layer west of the storm center. The calculations yielded wind-speed alterations of two or three miles per hour. In a few locations, though, the velocities changed by as much as 20 mph because of minor redirections of the winds near the storm's center.

Although the original and altered versions of Hurricane Iniki looked nearly identical in structure, the changes in the key variables were large enough that the latter veered off to the west for the first six hours of the simulation and then traveled due north, so that Kauai escaped the storm's most damaging winds. The relatively small, artificial alterations to the storm's initial conditions had propagated through the complex set of nonlinear equations that simulated the storm to result in the desired relocation after six hours. This run gave us confidence that we were on the right path to determining the changes needed to modify real hurricanes

If it is true, as our results suggest, that small changes in the temperature in and around a hurricane can shift its path in a predictable direction or slow its winds, the question becomes, How can such perturbations be achieved? No one, of course, can alter the temperature throughout something as large as a hurricane instantaneously. It might be possible, however, to heat the air around a hurricane and thus adjust the temperature over time.

Our team plans to conduct experiments in which we will calculate the precise pattern and strength of atmospheric heating needed to moderate hurricane intensity or alter its track. Undoubtedly, the energy required to do so would be huge, but an array of earth-orbiting solar power stations could eventually be used to supply sufficient energy. These power-generating satellites might use giant mirrors to focus sunlight on solar cells and then beam the collected energy down to microwave receivers on the ground. Current designs for space solar power stations would radiate microwaves at frequencies that pass through the atmosphere without heating it, so as to not waste energy. For weather control, however, tuning the microwave downlink to frequencies better absorbed by water vapor could heat different levels in the atmosphere as desired.

Another potential method to modify severe tropical storms would be to directly limit the availability of energy by coating the ocean surface with a thin film of a biodegradable oil that slows evaporation. Hurricanes might also be influenced by introducing gradual modifications days in advance of their approach and thousands of miles away from their eventual targets. By altering air pressure, these efforts might stimulate changes in the large-scale wind patterns at the jet-stream level, which can have major effects on a hurricane's intensity and track. Further, it is possible that relatively minor alterations to our normal activities—such as directing aircraft flight plans to precisely position contrails and thus increase cloud cover or varying crop irrigation practices to enhance or decrease evaporation—might generate the appropriate starting alterations.

If meteorological control does turn out to work at some point in the future, it would raise serious political problems. What if intervention causes a hurricane to damage another country's territory? And, although the use of weather modification as a weapon was banned by a United Nations Convention in the late 1970s, some countries might be tempted.

Have some countries or terrorist organizations, as Secretary of Defense Cohen suggested, been more than just tempted to use weather modification as a weapon? Has this technology passed the boundary from science fiction to science fact? Has the United States itself used weather as a weapon? The answer to that last question is a resounding "yes!"

PROJECT POPEYE

As we have seen, the American military-industrial-academic complex early on recognized the importance of weather as a weapon. After the great battles of the Civil War it was noted that rains seems to follow, perhaps caused by the smoke or concussions. A General patented an idea for making rain from this observation, but it would take nearly eighty years for a technology to be developed that was GI friendly. The Battle for Britain was partially won because Allied forces successfully used a fog-dispersal system known as FIDO to enable aircraft takeoff and landing under otherwise debilitating fog conditions. Cold fogs were similarly dissipated during the Korean War. Cloud seeding became a weapon in Vietnam under *Project Popeye.*

Project Popeye is a now exposed and proven conspiracy on the part of the military to circumvent the laws of humanity in time of war using environmental modification as a weapon — and to keep this secret the Secretary of Defense was forced to lie to Congress!

Project Popeye was originally conducted as a pilot program in 1966. It was an attempt to extend the monsoon season in Southeast Asia with the goal of slowing traffic on the Ho Chi Minh Trail by seeding clouds above it in hopes of producing impassable mud. Over the course of the program silver iodide was dispersed from C-130s, F4 Phantoms (a fighter jet) and the Douglas A-1E Skyraider (a single engine propeller driven fighter-bomber), into clouds over portions of the trail winding from North Vietnam through Laos and Cambodia into South Vietnam. Positive results from the initial test led to continued operations from 1967 through 1972.

Some scientists believe that it did hamper North Vietnamese operations, even though the effectiveness of this program is still in dispute. In 1978, after the efforts at cloud seeding in Vietnam produced mixed results, the U.S. Air Force declared its position to be that "weather modification has little utility as a weapon of war." Many have remarked on what obvious malarkey that statement was.

Recent military publications indeed have stated quite the opposite. For example the U.S. Air Force's own Air University's *SPACECAST 2020* contained a section on Counterforce Weather Control for force enhancement, which pointed out that:

> Atmospheric scientists have pursued terrestrial weather modification in earnest since the 1940s, but have made little progress because of scientific, legal, and social concerns, as well as certain controls at various

government levels. Using environmental modification techniques to destroy, damage, or injure another state are prohibited. However, space presents us with a new arena, technology provides new opportunities, and our conception of future capabilities compels a reexamination of this sensitive and potentially risky topic.

SPACECAST 2020 has been superseded by the previously mentioned *Air Force 2025,* which made this same point saying:

> The influence of the weather on military operations has long been recognized. During World War II, Eisenhower said, "In Europe bad weather is the worst enemy of the air [operations]. Some soldier once said, 'The weather is always neutral.' Nothing could be more untrue. Bad weather is obviously the enemy of the side that seeks to launch projects requiring good weather, or of the side possessing great assets, such as strong air forces, which depend upon good weather for effective operations. If really bad weather should endure permanently, the Nazi would need nothing else to defend the Normandy coast!"

Clearly, weather control could have a marked effect on the outcome of military operations. The problem the military has is not whether weather control should be affected, but how it could be done, meaning technically, legally and politically. Many researchers, myself included, believe that the DoD never truly gave up trying to find out.

The existence of Project Popeye may have first come to light when Dr. Daniel Ellsberg released the so-called *Pentagon Papers* in 1970. *Wikipedia,* the free online encyclopedia tells us:

> The Pentagon Papers is the colloquial term for "United States-Vietnam Relations, 1945-1967: A Study

Prepared by the Department of Defense," a 47 volume, 7,000-page, top-secret United States Department of Defense history of the United States' political and military involvement in the Vietnam War from 1945 to 1971, with a focus on the internal planning and policy decisions within the U.S. Government. The study was commissioned in 1967 by Robert McNamara, the then-Secretary of Defense. The Papers included 4,000 pages of actual documents from the 1945-1967 period, and 3,000 pages of analysis.

Project Popeye reached broad public consciousness when syndicated columnist Jack Anderson revealed it under the code name *Intermediary-Compatriot* in his *Washington Post* column of 18 March 1971.

US Defense Secretary Melvin Laird was forced to testify before Congress about it in 1972. He told the US Senate that Anderson's wild tales were untrue and that the United States never tried to seed clouds in Southeast Asia. But on 28 January 1974 a private letter from Laird was leaked to the press. By 1974 he had left Defense and was counsel to President Nixon who was fighting for his political life following the break-in at the Democratic Party's National Committee offices in the Watergate Hotel on 17 June 1972. In the letter he privately admitted that his 1972 testimony had been false and that the US did in fact use weather modification in North Vietnam in 1967-68.

On 20 March 1974 the United States Senate held a top secret hearing in which representatives of the military finally admitted to the existence of Operation Popeye. They conceded that the cloud seeding program had been conducted over neutral Cambodia and Laos (in violation of international law), as well as both North and South Vietnam. The testifying Pentagon officials stated that Popeye had been ongoing from 1966 through 1972 and that at least 2,600 flights had released over 47,000 units of cloud-seeding materials during the program, at a total cost for the operation of around $21.6

million.

These hearings also revealed that the US military had attempted other environmental modifications as well. The US had used massive spraying of chemical herbicides in the hopes of depriving its foes of both food supplies and shelter. The US strategy was developed to counter Viet Cong (Vietnamese National Liberation Front) guerrilla tactics inspired by Mao Tse-Tung. Mao had advocated use of hidden bases and unpredictable attacks to maintain guerrilla initiative (the very tactics the armed forces of the British American colonies had used two hundred years earlier to win their freedom from British rule to become The United States of America!).

The US attempted environmental modification in order to make the Southeast Asian environment serve U.S. needs rather than those of the Viet Cong. According to analyst L. Juda (from "Negotiating A Treaty On Environmental Modification Warfare: The Convention On Environmental Warfare And Its Impact On The Arms Control Negotiations," published in *International Organization*) the idea was simple:

> If, as has been suggested, then the guerrilla is to his base area as fish are to the sea, the destruction of the sea would kill the fish and the elimination of the base area with its supports would destroy the guerrilla.

The implications of this operation staggered Senator Claiborne Pell, a Democrat from Rhode Island. In 1976 he said:

> The U.S. and other world Powers should sign a treaty to outlaw the tampering with weather as an instrument of war. It may seem farfetched to think of using weather as a weapon—but I am convinced that the U.S. did in fact use rainmaking techniques as a weapon of war in Southeast Asia. We need a treaty now to prevent such actions—before military leaders of the world start directing storms, manipulating climates

and inducing earthquakes against their enemies. It may seem a great leap of imagination to move from an apparent effort by the United States to muddy the Ho Chi Minh trail in Laos by weather modification to such science fiction ideas as unleashing earthquakes, melting the polar ice cap, changing the course of warm ocean currents, or modifying the weather of an adversary's farm belt. But in military technology, today's science fiction is tomorrow's strategic reality.

Of course, he wasn't the first to see the handwriting on that wall. A prescient article in *Fortune* magazine in February of 1948 had spelled out the horrible reality of where EnMod could go, with:

> Army, Navy, and Air Force are spending close to a million dollars a year on weather modification and their tremendous interest suggests that military applications extend far beyond visiting a few showers upon an enemy. It does not require a sharp mind to figure out that wartime storms might readily be infected with virulent bacteriological and radiological substances.

Senator Pell had conducted the Senate hearing in 1972 in which he was lied to by Defense Secretary Laird and the secret one in 1974 that learned the horrible truth. After these he became a leading advocate for what became the EnMod Convention. A subcommittee chaired by Minnesota Congressman Donald Fraser did the same in the House of Representatives in 1974 and 1975. The treaty to ban all hostile uses of environmental modification techniques was proposed, and warmly received by the international community, in no small part as a response to revelation that the United States' military had used environmental modification techniques during the Vietnam War.

The 1978 Convention on the Prohibition of Military or Any

Weather Warfare

Other Hostile Use of Environmental Modification Techniques (EnMod Convention) prohibits the use of techniques that would have widespread, long-lasting or severe effects through deliberate manipulation of natural processes and causing such phenomena as earthquakes, tidal waves and changes in climate and in weather patterns.

Before we examine the EnMod Convention in detail, let us first look at other events that may have led to its signing.

CHAPTER THREE

EARTHQUAKES
ON DEMAND?

Officially, there is an area of research devoted to man-made earthquakes. Geologists and seismologists agree earthquakes can be induced in five major ways: fluid injection into the Earth, fluid extraction from the Earth, mining or quarrying, nuclear testing and through the construction of dams and reservoirs.

Geologists discovered that disposal of waste fluids by means of injecting them deep into the Earth could trigger earthquakes after a series of quakes in the Denver area occurred from 1962-1965; the periods and amounts of injected waste coincided with the frequency and magnitude of quakes in the Denver area. The earthquakes were triggered because the liquid, which was injected under very high pressure, released stored strain energy in the rocks.

So wrote Jason Jeffrey in his article "Earthquakes: Natural or Man-Made?" in an issue of *New Dawn Magazine* that appeared shortly after the Christmas 2004 Asian tsunami. Accidental causation of earthquakes is one thing, but the intentional triggering of a temblor by the military is another — or is it?

At first blush one is tempted to dismiss the concept of "earthquakes on demand" as simple lunacy. Such a gut reaction, however, is probably not consistent with scientific facts. Scientists around the world have indeed recognized and documented several mechanisms by which human activities have resulted in earthquakes. These mechanisms

61

were discovered by accident, it is true. But, how difficult is it to replicate the conditions that caused the initial accidental earthquake? If we can do it once by mistake, couldn't we do it again intentionally?

There have been officially recorded instances of earthquakes caused by all five human activities mentioned by Jason Jeffrey above. Scientists from the Institute of Dynamics of Geospheres at the Russian Academy of Sciences have also observed that earthquakes can be triggered by human action. They report:

> Induced seismicity, or seismic activity caused directly by human involvement, has been detected as a result of water filling large surface reservoirs, development of mineral, geothermal and hydrocarbon resources, waste injection, underground nuclear explosions and large-scale construction projects. If the stress change is big enough, it can cause an earthquake, either by fracturing the rock mass—in the case of mining or underground explosions—or by causing rock to slip along existing zones of weakness.

The China Yangtze Three Gorges Project Development Corporation, which is building the controversial Three Gorges Dam on the Yangtze River in China, has been concerned about earthquakes since the inception of the project. The Three Gorges project will be the world's largest dam complex. It has been engineered to store over 5 trillion gallons of water and to withstand an earthquake of 7.0 on the Richter scale—based in part on fears that the sheer weight of the dam and all that water could cause an earthquake that big.

Underground nuclear explosions have also set off earthquakes. The best documented of these occurred deep in the vast Nevada desert in 1968. The United States government set off a nuclear weapon on a fault line to see if they could induce an earthquake—and got one! This took place under the code name *Project Faultless*.

PROJECT FAULTLESS

According to the Nevada Division of Environmental Protection *Project Faultless* was detonated at the Central Nevada Test Area (CNTA) in emplacement borehole UC-1 on 19 January 1968 at a depth of 3,200 feet below ground surface. Officially, Faultless was designed to study the behavior and characteristics of seismic signals generated by nuclear detonations and to differentiate them from seismic signals generated by naturally occurring earthquakes, and also to evaluate the usefulness of the site for higher-yield nuclear tests that could not be safely conducted at the Nevada Test Site (NTS) near Las Vegas. It was intentionally placed on a fault line to see if the high yield explosion would set off an earthquake. It has been argued that the scientists in charge of the shot were convinced that such was impossible, hence the name "faultless."

Michon Mackedon teaches English and Humanities at Western Nevada Community College in Fallon, Nevada. She has served as Vice Chairman of the Nevada Commission on Nuclear Projects since 1986 and is writing a book about nuclear projects in Nevada. She tells us the story of this successfully induced, albeit seemingly unintentional earthquake in her article "Project Faultless: Central Nevada's Near Miss as an Atomic Proving Ground," which can be found on the Eureka County Yucca Mountain Information Office website. As this is an official government website, owned by "we the people," Ms. Mackedon has graciously given this article to the world copyright free. As it is a hell of a story, including evidence of how the government lies to us (posted on a government site no less!) I'm going to include most of it here:

> The whole notion of testing nuclear bombs is startling, like testing tornadoes or simulating the "Big Bang." Nevertheless, the United States has been engaged in testing nuclear weapons since July 16,

1945, when the atomic test code-named Trinity was conducted in Alamogordo, New Mexico. On that day, in the predawn hours, the nucleus of an atom was split to create a weapon of such overwhelming power that all those witnessing the event were forever affected by what they had seen there.

It has been said that the Trinity test let the nuclear genie out of the bottle, but it learned to dance on the deserts of Nevada. In 1951, the Nevada Test Site (NTS), located 90 miles north of Las Vegas was selected to serve as the nation's continental atomic proving ground. Earlier, in 1946 and 1948, atomic bombs had been tested in the Marshall Islands, but weapons designers living in New Mexico wanted a proving ground closer to their laboratories, and the Korean War increased the pressure to improve atomic weapons designs.

Over the next four decades, The Nevada Test Site became home to 928 nuclear tests. (A testing moratorium halted nuclear testing in 1992). One hundred of the tests conducted there were atmospheric tests — dropped by aircraft or exploded from towers, balloons, or cannons — producing the signature mushroom cloud and dangerous radioactive fallout. In 1963, under the terms of a Limited Test Ban Treaty signed by President Kennedy and Russia's Kruschev, all the tests were moved underground. The remaining 828 NTS nuclear tests were conducted in shafts and tunnels.

After nuclear testing was moved underground, a series of problems at NTS drew the eyes of nuclear planners to a section of land in Nye County's Hot Creek Valley, approximately midway between Eureka and Tonopah, where plans were quietly laid to develop a supplemental nuclear proving ground.

One of the problems faced by test administrators was the growing tourist economy in Las Vegas. In the early years of atomic testing, the Las Vegas community was quite small, and most of the residents and tourists who visited there accommodated atomic testing as part of the overall Las Vegas experience. Visitors even scheduled special trips to the area during "bomb season" to drink "atomic cocktails" while they awaited the dramatic spectacle of an exploding atomic bomb.

The move to underground testing, while reducing fallout danger, produced ground motion which could be felt in Las Vegas. The larger the test the greater the ground motion: the swaying earth and occasional shattered glass window were making tourists, residents, and casino operators quite nervous. At the same time, in the mid 1960s, the military was developing a Spartan anti-ballistic missile, capable of carrying and delivering a multiple megaton warhead, and they planned to test the effectiveness of even larger nuclear weapons than had been tested before.

So, with some urgency, the search for a supplemental test site was launched. By 1965, the potential sites had been narrowed to Amchitka, Alaska and Central Nevada. However, before the Central Nevada site could be granted full status as a Proving Ground, the Atomic Energy Commission (AEC) wanted to determine how the geology of the area would respond to multi-megaton underground explosions. As a result, the first test scheduled there, and, as it happened, the only test to take place there, was categorized as a calibration test (as opposed to a weapons effect test, where the effects of the bomb on animals, homes, military equipment and bomb shelters were measured). The test was planned to yield just under one megaton, qualifying it as a very large test, yet the projected size of the event was never announced to the public.

In 1967, the AEC public relations teams headed to Eureka and Tonopah to prepare the residents for the calibration test, which was assigned the code-name Project Faultless and tentatively scheduled for early

1968. The AEC faced a larger than usual challenge in trying to convince residents of Central Nevada that the test would be beneficial to their communities, as the area slated for the calibration study lay in the fallout path of over half of the atmospheric tests of the 1950s. In particular, people there still felt the loss of eight year old Butch Bardoli, who had succumbed to leukemia in 1957, a disease which his family felt was directly related to atomic fallout over their Nye County ranch.

The AEC at first proposed to deal "openly" with the test, so as to head off the accusations of "secrecy" so often leveled at nuclear testing. From past experience they knew that closing off the test area would leave the whole endeavor vulnerable to wild speculation. The military, however, wanted to keep observers away from seeing the predicted 14-16 foot ground swell, which might alert them to size of the Faultless device. The compromise between the positions is, in retrospect, quite funny and serves as an indication of how the public relations arm of nuclear testing manipulated language to get the job done.

An AEC internal memo proposed closing "the line of sight" to the test but setting up cameras to televise images of the shot back to the base camp. The memo states, "observe does not mean 'see,' but rather be on hand at a selected area. ...the 'observer' site might be established with no line of sight to ground zero. The 'observer' area will be at the AEC base camp, about 30 miles from the base camp."

Another memo which seems retrospectively humorous discusses the sticky dilemma of keeping the public "informed" about the test without really giving out information which might lead to uncomfortable questioning or even protest. It reads, "Statements as to 'how soon' nuclear detonations can be arranged should be couched only in general terms. No readiness dates should be given...and it should be clearly explained that a readiness date is not a schedule...."

The public relations procedures for selling Project Faultless to the people of central Nevada are of interest to those who listen to the current debate over using Yucca Mountain as a high level nuclear waste dump. Among the techniques used to sell nuclear projects, then and now, are emphasizing necessity, stressing local benefits and downplaying risk.

A transcript of a Tonopah town meeting, called by the AEC three months in advance of the January 1968 test, discloses efforts made by the AEC's advance team to convince central Nevada residents to open their arms to nuclear tests. Necessity was emphasized by invoking the Cold War: "First, the United States [Atomic] Energy Commission [and] the Department of Defense has been able to maintain reasonable parity with the Soviet Union by conducting an aggressive and certainly expensive underground nuclear test program...."

The Local Benefits were emphasized by alluding to the economic gains netted in Fallon, Nevada, where a relatively small (12 kiloton) device had been detonated underground in 1963 in a test code-named Project Shoal: "We think we have interfered as little as possible with their way of life, and we sincerely hope that we have contributed something to their economy..... We need places for our engineers; we need places for our technicians; we need fuel for our vehicles.... A large amount of money and effort will be expended in studying the water area systems of the area..."

As for the risk, those in attendance were told, "The worst it can do to the public is inconvenience [them]."

The extent to which the potential economic boon to the area became a selling point for the project is underscored by a letter sent by a central Nevada rancher to Nevada Senator Alan Bible in late 1967. He wrote:

> *A lot has been said by the AEC spokesmen and representatives of the Government about the economic benefits to the local communities, when the AEC moves in men and equipment. Indeed I understand that different states exerted all effort and influence to get the test sites located in them, because of the tremendous amount of money this program would bring in.*
>
> *This kind of temporary prosperity is hollow and false and is promoted by people who can see no farther than their cash register. It is prosperity for a few local merchants and gamblers, at the expense of the state's lands and water resources, land and water, that if they are not damaged, will produce far more real prosperity than this testing program.*

As the preparations for Project Faultless proceeded, the legendary Howard Hughes stepped into the picture to add more pressure on the AEC to move megaton testing away from Las Vegas to central Nevada. Mr. Hughes was a walking paradox—a mysterious hermit and well-known casino mogul, whose odd habits and immense bank account made him a force to reckon with in Las Vegas politics and atomic planning. Hughes became convinced that the increasingly large underground tests at NTS would ruin the Vegas gaming industry. He became quite vocal in his opposition to megaton testing at NTS, and his organization even tried to delay one of the larger NTS tests. He began to pressure AEC to move all underground testing away from Las Vegas to the proposed site in central Nevada or to the supplemental site being studied in Amchitka, Alaska.

A personal memo kept by Test Site manager

J.T. Reeves reveals what the situation was in early 1968. He noted that Howard Hughes had called him "approaching hysteria" over the Boxcar test planned at NTS for the spring of 1968. Hughes was "under the impression" that the AEC had informed its people that they were going to discontinue tests at NTS and move future testing to central Nevada and Alaska. Hughes told Reeves that he was not about to "invest money in the development of a new airport in Clark County with the AEC continually permanently damaging buildings in the area and contaminating the atmosphere and ground water...."

Much was riding on the success of the Faultless Test. Three deep emplacement holes were drilled on the Central Nevada site, one for Faultless and the other two in anticipation of tests which would immediately follow a successful calibration of the site. A second test was even assigned the code-name, Adagio. The size of the Adagio drillhole suggests that it was planned to be a test in the multi-megaton range, three or perhaps 4 megatons. (The largest underground test ever conducted by the U.S. was 5 megaton Cannikin, in 1971, at Amchitka, Alaska.)

On January 19, 1968, the Faultless Test commenced, with observers from the public stationed, as planned, out of the line of sight. What they would have witnessed was a dramatic fifteen foot upheaval of the earth above ground zero. Then, the earth collapsed north and south of ground zero, leaving massive fault blocks extending for thousands of feet. In some places the drops measured 10 feet. Eighty-seven miles away from the explosion, windows broke at White Pine High School in Ely.

The surface damage was dramatic proof that the experiment to calibrate the site had overreached itself. The Central Nevada Test Area was eventually declared unsuitable for further underground nuclear tests.

Adagio was cancelled, and Hughes was distraught.

There are two epilogues to the Faultless story. One is the resolution of the Hughes dilemma. In about May of 1969, a year following the Faultless failure, Hughes began to withdraw his objections to testing at NTS. John Meier of the Hughes Organization visited Robert Miller of the AEC Nevada Operations Office saying that there had been confusion and "division of opinion" within the Hughes Tool Company regarding nuclear testing. Meier "admitted the paradox that it was clearly in their interests from a financial point of view to support the administration's [nuclear anti-ballistic missile] position." The Hughes Company was ready to do business again with NTS. The last note in the Howard Hughes file, now held in DOE archives in Las Vegas, is a letter sent to AEC's Miller from the Hughes Tool Company on January 31, 1972: "We wish to take this opportunity in starting another year to thank you for your business and support. All of us at Hughes are grateful for the privilege of serving you."

Marketplace economics had apparently resolved the issues.

The second epilogue is of more contemporary interest and concerns the future of the Central Nevada Test Area. When an underground nuclear test takes place, a huge subterranean cavern is formed, trapping within it intensely radioactive particles. The surface then collapses into the cavity creating what is called a rubble chimney, a funnel of debris linking the surface and the cavity. Over time, ground water will seep into the chimney and cavity, allowing the radioactive particles to migrate away from ground zero toward populations. Many of the radionuclides remaining from an underground nuclear test are extremely dangerous and very long-lived. Plutonium, the most dangerous residue, has a half-life of 24,500 years.

The Faultless site is now marked by a bronze

tablet attached to a steel pipe protruding from the ground above the cavity. The marker posts restrictions on drilling, but does not provide reasons for the restrictions, nor does it notify the public that it will be thousands of years before the radioactive materials decay. Who will carry the story forward through eons of time and warn others of the atomic danger beneath the ground?

If there are lessons to be learned from past mistakes, then the legacy left by Project Faultless and other nuclear experiments should serve as warning to those committed to emplacing high level radioactive waste in Yucca Mountain. Buried radionuclides, whether the result of a nuclear explosion or a nuclear repository, will remain dangerous through dozens of millenniums, posing a threat to water supplies essentially forever.

Another lesson from the past is the uncertainty involved in predicting the performance of Mother Earth. The code-name assigned to Faultless reflects the "scientifically-based" prediction made by the planners of the test that the site would prove geologically stable, literally fault-less, under the pressure of a megaton blast. The failure of that prediction colors the code-name with unintended irony and suggests that prediction is risky business, especially when the stakes are as high as they are when we deal with dangerous radionuclides. Yet Yucca Mountain planners continue to predict that the repository will not be disturbed by earthquakes, miners or curiosity seekers, upwelling or downpouring water, or anything else, for at least 10,000 years.

As a twenty-year Nevada resident and a lifelong opponent of nuclear power I have to admit to you that I do oppose the plan to place a high-level nuclear repository in Yucca Mountain—but the details of my opposition will have to wait for another book, I'm afraid.

Project Faultless was hardly the only nuke shot to produce an earthquake—but it was the best documented one. On 19 June 1992 the US conducted an underground atom bomb test at the Nevada Test Site (NTS) which was followed by another test just four days later. A series of heavy earthquakes as high as 7.6 on the Richter scale rocked the Mojave Desert 176 miles to the south three days after the second test. Only 22 hours after that an "unrelated" earthquake of 5.6 struck less than 20 miles from the NTS! It was the biggest earthquake ever recorded near the test site. It caused one million dollars in damage to buildings at Yucca Mountain. The Yucca Mountain facility was only fifteen miles from the epicenter of the earthquake!

In a statement issued on 14 July 1992 responding to the "understandable unease" expressed by the public, the Department of Energy in Washington (having conveniently forgotten Project Faultless) asserted the relationship between nuclear testing and earthquakes was "nonexistent."

Among those frightened by nuclear testing was one Estes Kefauver. In 1956 he was the Democratic Vice-Presidential candidate, running mate to Adlai Stevenson. Adlai Ewing Stevenson II (5 February 1900 – 14 July 1965) was a one-term governor of Illinois who lost by landslides in two races for president against Dwight D. Eisenhower in 1952 and 1956. While on the campaign trail Kefauver warned: "H-bomb tests could knock the earth 16 degrees off its axis!"

His warning became the basis for one of the greatest sci-fi films to come from England, 1961's *The Day the Earth Caught Fire*. Fiona Kelleghan wrote this summary of it for the Internet Movie Database:

> Journalists of the *London Daily Express* investigate reports of strange phenomena occurring all over the world, such as flooding in the Sahara, unseasonable blizzards in New York, and violent tornadoes in the Soviet Union. All over England, temperatures are on the rise, girls in bikinis are everywhere, and wonderful

special-effects mists are blanketing the Thames River. Top scientists at the Meteorological Center refuse to give any official explanation, which makes the newspaper editor suspicious. He orders science reporter Bill Maguire (Leo McKern) and alcoholic columnist Peter Stenning (Edward Judd) to dig for information. When Peter begins a romance with Met Center secretary Jeannie Craig (Janet Munro), he learns from her certain clues that there has indeed been a cover-up... and he begins to sober up, so that he may win her love. Ten days before the film begins, two nuclear bombs had been exploded, one at the North Pole by the Soviet Union, the other at the South Pole by the United States. Nobody noticed before now that they were set almost simultaneously, until the Chief Editor, in a conference with his reporters, draws a line from London to New Zealand, showing the path that floods and other catastrophes have created in a devastating line. Stenning gathers from hints Jeannie gives him that the two explosions shifted the Earth's orbit and set it on a course toward the sun. As water becomes scarce, the British government takes emergency measures to control hysteria, rampant looting, and rioting by teenagers. The only possible solution is for another explosion of bombs that will restore the Earth's orbit. On detonation day, Stenning heroically makes his way through a wasteland to the newsroom to write a story that will prepare for the results of the blasts. He instructs the typesetters to prepare two front pages. One has the headline "World Saved," the other "World Doomed."

A study conducted in 1975-76 by two Japanese scientists, entitled "Recent Abnormal Phenomena on Earth and Atomic Power Tests," by Shigeyoshi Matsumae, President of Tokai University, and Yoshio Kato, head of the University's Department of Aerospace Science, concluded:

> Abnormal meteorological phenomena, earth-quakes and fluctuations of the earth's axis are related in a direct cause-and-effect to testing of nuclear devices... Nuclear testing is the cause of abnormal polar motion of the earth. By applying the dates of nuclear tests with a force of more than 150 kilotons, we found it obvious that the position of the pole slid radically at the time of the nuclear explosion... Some of the sudden changes measured up to one meter in distance.

Not quite Kefauver's 16 degrees off the axis, but scary enough. Two years later, on 12 October 1978, the British magazine *New Scientist* reported:

> Geophysicists in Germany and England believe the 1978 earthquake in Tabas, Iran, in which at least twenty-five thousand people were killed, may have been triggered by an underground nuclear explosion. ...British seismologists believe the Tabas earthquake implies a nuclear test that has gone awry. ...Moreover, a seismic laboratory in Uppsala, Sweden, recorded a Soviet nuclear test of unusual size—ten megatons—at Semipalitinsk only thirty-six hours before. ...One German scientist specifically implicated this test in the origin of Tabas disaster.

The *War and Peace Digest* is a bimonthly international newsletter on issues of disarmament, government secrecy, media accountability, the nuclear threat (from both civilian power plants and the military weapons complex), ecological destruction, and peaceful conflict resolution through the structures of the United Nations. The August 1992 edition featured an article that reported on a presentation delivered on 14 April 1989 to the Second Annual Conference on the United Nations and World Peace in Seattle, Washington, by Gary T. Whiteford, Professor of Geography at the University of New Brunswick in Canada. Professor Whiteford presented an exhaustive study of the correlations between nuclear testing and earthquakes in a paper entitled: "Earthquakes and Nuclear Testing: Dangerous Patterns and Trends." *War and Peace Digest* told its readers:

> Whiteford studied all earthquakes this century of more than 5.8 on the Richter scale. "Below that intensity," he explained, "some earthquakes would have passed unrecorded in the earlier part of the century when measuring devices were less sensitive and less ubiquitous. But for bigger quakes the records are detailed and complete for the entire planet." So Whiteford was able to make a simple comparison of the earthquake rate in the first half of the century, before nuclear testing, and the rate for 1950 to 1988. In the fifty years before testing, large earthquakes of more than 5.8 occurred at an average rate of 68 per year. With the advent of testing the rate rose "suddenly and dramatically" to an average of 127 a year. The earthquake rate has almost doubled. To this day the U.S. military attributes the increase to "coincidence." As Whiteford comments, "The geographical patterns in the data, with a clustering of earthquakes in specific regions matched to specific test dates and sites do not support the easy and comforting explanation of `pure coincidence.' It is a dangerous coincidence."
>
> Within the data he found other suggestive patterns. The one-two nuclear test punch that preceded by only a few days the July earthquakes in California this year may reveal a special danger. The largest earthquake this century took place in Tangshan in North-East China on July 27 1976. It measured 8.2 and killed 800,000 people. Only five days earlier the French had tested a bomb in the Mururoa atoll in the Pacific. Four days later the United States tested a bomb in Nevada. Twenty-four hours later the earthquake hit China.

75

In an even more revealing analysis, Whiteford studies so-called "killer earthquakes" in which more than one thousand people have died. He compiled a list of all such quakes since 1953 and matched them with nuclear test schedules. Some test dates were not available, but in those that were, a pattern was evident: 62.5% of the killer earthquakes occurred only a few days after a nuclear test. Many struck only one day after a detonation. More than a million people have now died in earthquakes that seem to be related to nuclear tests. Again, the governments of the nuclear nations claim the results are mere coincidence. Officially the U.S. energy department maintains that even their most powerful nuclear tests have no impact beyond a radius of 15 miles. The claim is challenged by the instruments of modern seismology that can register nuclear tests anywhere in the world by measuring local geological disruptions. Whiteford speculated that although the reverberations may fade within fifteen miles of a test, they are merely the first ripple of a wave that travels through the planet's crust and spreads around the globe.

In 1991 the Nuclear Age Peace Foundation published Whiteford's findings in an article called "Is Nuclear Testing Triggering Earthquakes and Volcanic Activity?" In an interview with California State seismologist, Dr. Lalliana Mualchin, the foundation went on to inquire into the long-term effects of testing. Mualchin was asked if the cumulative effect of nuclear testing might be to trigger earthquakes and volcanoes.

He replied, "A single nuclear test may have little effect on the earth, like that of an insect biting an elephant. But the cumulative effect might move the earth's tectonic plates in a manner similar to how a swarm of insects might start an elephant running."

Mualchin added, "If an insect bites an elephant in

a sensitive spot, such as an eye or an ear, then there might be a vast movement out of all proportion to the size of the bite."

The article concluded, "Who will the world hold responsible if suddenly an unprecedented series of violent earthquakes and volcanoes shake the earth? Will nuclear testers be able to assure the world they were not responsible?"

That last bit is a haunting reminder of the legal problems with civilian weather modification, is it not? If we have such problems when the damage occurs in the open in the civilian sector, how much more difficult will it be to hold responsible the military and their governments who deny it ever happened at all?

Also, that "If an insect bites an elephant in a sensitive spot, such as an eye or an ear, then there might be a vast movement out of all proportion to the size of the bite" is an stunning analogy for what Dr. Gordon J. F. MacDonald meant by: "The key to geophysical warfare is the identification of environmental instabilities to which the addition of a small amount of energy would release vastly greater amounts of energy."

Human activity, both military and civilian, has been proven capable of generating earthquakes. If earthquakes can be brought about by accident then it follows that they can be intentionally generated as well. It would seem that Senator Pell and Secretary Cohen were right to be worried!

ELECTROMAGNETIC WAVES AND EARTHQUAKES

What about Secretary Cohen's contention that eco-terrorists may be able to produce earthquakes using electromagnetic (radio) waves? Is it really possible?

The Washington Times reported 29 March 1992 on the incidence of extremely low frequency (ELF) radio signals being associated with earthquakes, stating: "Satellites and ground sensors detected mysterious radio waves or related electrical magnetic activity before major earthquakes in Southern California during 1986-87, Armenia in 1988, and Japan and Northern California in 1989." An Athens University physicist was also reported to have observed electromagnetic signals in six out of seven quakes in Greece over several years. Were these quakes generating the radio signals, or was it the other way around?

Let us first consider the possibility that the signals are generated in the earth by the impending earthquake. Perhaps you've heard of "Charlotte's Syndrome"…

The bizarre case of Charlotte King, who "hears" the Earth, yields valuable insights into perception, physics, and the geosciences. Charlotte King was taped by the *Good Evening* crew (KGW-TV, Portland), giving warning before the 22 April 1992 Joshua Tree magnitude (M) 6.3 shaker; then, she predicted the Cape Mendocino earthquakes of Saturday and Sunday, 25 April (M7.1) and 26 April (M6.6; M6.7). On Monday she warned that, "It's not over yet," for Southern California. The tape aired on Monday, April 27, 1992. As we'll see in a moment, the Landers and Big Bear quakes hit two months later on 28 June.

When a volcanic eruption started nearly anywhere in the world Charlotte King felt abdominal pain. With Mt. Saint Helens, the strong pains doubled her over. When the mountain erupted on 18 May 1980 she suffered a minor stroke. This was reported in her 1981 appearance on the TV show *That's Incredible!*

One scientific research team spent fourteen years investigating "Charlotte's Syndrome." They concluded that she was sensitive to fluctuations in the earth's electromagnetic field. Piezoelectricity is a little understood phenomenon. Scientists have demonstrated that under certain conditions crystals can be forced to give off a flow of electricity if subjected to high pressure. The earthquake precursor signals Charlotte King feels could be the result of tectonically induced piezoelectricity. That is, pressure in the earth's crust could

78

squeeze some types of rocks so hard that they discharge electricity. The electrical flow would have electromagnetic effects, perhaps including a "broadcast" in the ELF range.

Several researchers in the UFO field have discovered that some UFOs might actually be "earthquake lights" caused by this same piezoelectric phenomena. Lights leaping away from mountain peaks have been seen and recorded since ancient times, and could be a similar type of discharge.

Professor Elizabeth A. Rauscher received her B.S., M.S. and Ph.D. from the University of California, Berkeley, in Nuclear and Astrophysics. She was a staff researcher at Lawrence Berkeley National Laboratory, Berkeley, for over nineteen years and taught at Berkeley, Stanford, J.F. Kennedy University and the University of Nevada. She has also worked on the NASA space shuttle program and has been a delegate to the United Nations on long-term energy sources. She holds three US patents with Dr. William L. Van Bise. The meaning of life, existence, truth and reality, and the nature of consciousness are her main interests. She has researched psychic phenomena and psychic healing for over thirty years and has published four books and over 250 papers.

Dr. Rauscher and Dr. William L. Van Bise, Sc.D., an electrical engineer, called the Earthquake Prediction Registry at the Library of Congress on 8 January 1994 to report impending events likely to occur within 30 days. Unique signals from the earth indicated that one or more quakes would occur in or near the Los Angeles area. The Northridge quake struck nine days later on 17 January. Unusual surges of signals from 3.8 to 4.0 Hz were recorded beginning two weeks before the quake.

Previously, Drs. Rauscher and Van Bise had attended an International Workshop held at Lake Arrowhead, California, 14-17 June 1992 with the title: "Low Frequency Electrical Precursors: Fact or Fiction?" Rauscher and Van Bise presented a paper on measurements of ELF signals. Rauscher announced that a magnitude 7 or greater earthquake would strike "in the region of the conference, very soon." On 28

Weather Warfare

June the Landers quake, (M7.5), struck 44 miles east of the conference site; several hours later Big Bear Lake, only 20 miles east, was hit by a M6.6 temblor. They were able to focus on the area, timing, and strength because of extensive contacts with Charlotte King. They used an antenna array located near Reno, Nevada, to pick up signals at 3.8 cycles per second.

ELF waves may be a natural component of earthquake phenomena. The question then is, by mimicking natural forces or by exploiting weaknesses in nature as Dr. MacDonald suggested, could the military, rogue states or terrorists cause earthquakes by generating the right sort of ELF waves? Before you decide that question, let's look at other types of waves and what they may be able to do.

The five or six officially recognized sources for human caused earthquakes are not the only ones, as you may have guessed. One of the earliest validated man-made earthquakes happened in New York City in the 1890s!

TESLA'S EARTHQUAKE MACHINE

Nikola Tesla (10 July 1856 – 7 January 1943) was a world-renowned Serbian-American inventor, physicist, mechanical engineer, and electrical engineer. Many regard him as one of the most important inventors in human history; certainly he was one of the greatest geniuses of the 19th century. Among his lesser-know "accomplishments" was that he nearly flattened New York City in the 1890s with an artificially created earthquake. He called his technology for transmitting mechanical energy through the earth "TeleGeoDynamics." He was also famous, or infamous, for his experiments in the wireless transmission of electrical energy—free broadcast power.

Tesla was born in the town of Smiljan, in what was then the Austro-Hungarian Empire, to a family of Serbian descent. His father was an Orthodox priest and his mother, while unschooled, was highly intelligent and something

80

of an inventor herself. Young Nikola was a dreamer with a poetic touch. As he matured he developed the quality of self-discipline and a mania for precision. In old age he added a remarkable catalog of phobias and weird behavior to his personality.

From an early age he was fascinated with machines and science. He attended the Technical University at Graz, Austria, and the University of Prague, intending to pursue an engineering career. It was at Graz that he first saw the Gramme dynamo, which worked as a generator in one direction and, when reversed, became an electric motor in the other. Seeing this set him to thinking of ways to turn alternating current to advantage. Later, while visiting Budapest, he visualized the principle of the rotating magnetic field and began working up plans for an induction motor based on his vision.

In 1882 Tesla traveled to Paris, France, where he found employment at the Continental Edison Company. The following year, while on assignment to Strasburg, he used his spare time to construct his first induction motor. It was a great technological achievement, though few around him grasped that at the time. His attempts to interest his European contemporaries in his work proved disappointing at best. Going into the 1880's direct current (DC) was all the rage, while alternating current (AC) power was discredited as being unsafe and unworkable. Annoyed with the know-it-all smirking of people he considered fools, Tesla sailed for America in 1884.

The cocky twenty-seven-year-old arrived in New York with four cents in his pocket, a few of his own poems, and calculations for a flying machine (which he patented in 1928). He had little difficulty securing a position in Thomas Edison's organization. Unfortunately, the two inventors were far apart in background, temperament and methods. Their separation, on decidedly less than friendly terms, was inevitable. In time their intense competition in business and scientific discoveries drew newspaper headlines. Each despised the other. Edison thought Tesla was a swaggering snob. Tesla, who saw himself as a real scientist, considered Edison a mere "tinkerer," and a publicity hound at that.

For the record, *Wikipedia* tells us:

> Thomas Alva Edison (February 11, 1847 –October 18, 1931) was an American inventor and businessman who developed many devices which greatly influenced life in the 20th century. Dubbed "The Wizard of Menlo Park" by a newspaper reporter, he was one of the first inventors to apply the principles of mass production to the process of invention, and can therefore be credited with the creation of the first industrial research laboratory. Some of the inventions credited to him were not completely original, but improvements of earlier inventions, or were actually created by numerous employees working under his direction. Nevertheless, Edison is considered one of the most prolific inventors in history, holding 1,097 U.S. patents in his name, as well as many patents in the United Kingdom, France, and Germany.

Tesla soon established his own laboratory. Between 1886 and 1898 he received grants for 85 patents from his work there, with more still to come. Of the 46 basic patents in alternating current, Tesla eventually came to hold 45. Besides working out the many systems and machines needed to realize an alternating current technology, his interests took him into investigating a wide range of other scientific possibilities.

He experimented with shadowgraphs similar to those that would later be used by Wilhelm Konrad Roentgen (1845 – 1923) when he discovered X-rays in 1895 (for which Roentgen received the Nobel prize in 1901). Tesla's countless experiments included work on a carbon button lamp (two patents granted in 1886), on the power of electrical resonance, and on various types of lighting (for which three patents were granted to Tesla in 1891 alone). He worked with globes and tubes filled with gases that gave off light when electrically excited. This led to the creation of fluorescent lighting, so popular in office buildings today. Indeed, his laboratory was brilliantly illuminated with these lamps well before a long since forgotten assistant at Edison's lab invented the incandescent bulb—for which the publicity hungry Edison lost no time in taking full credit.

In the 1880s and '90s Tesla also worked with radio-frequency electromagnetic waves. Despite the claims made by Guglielmo Marconi (1874 – 1937), it was actually Tesla who did much of the most basic pioneering work in radio frequency technology. A United States Supreme Court decision overruled the Marconi patent, awarding it to Tesla. The Supreme Court concluded that Marconi had pirated Tesla's invention, using 14 of his patents. It was a classic case of industrial espionage. Marconi had begged Tesla to let him work in his lab and then repaid Tesla's kindness by stealing his work!

Tesla also theorized the ability to locate objects in the air or in the ground by using radio waves. Today we call Tesla's idea for finding things in the air with radio waves RADAR (RAdio Detecting And Ranging, a term coined by Captain S. M. Tucker USN in the 1930s).

In 1898 two years before Marconi stunned the world with his first wireless telegraph broadcast, Tesla announced his invention of a "teleautomatic" boat. When skepticism was voiced, Tesla proved his claims for radio remote control by demonstrating it before a crowd in Madison Square Garden with a radio controlled boat clearly demonstrating the lack of wires to guide the craft. Tesla might have been remembered as the Father Of Radio had not his ambitions run to a far more grandiose scheme—free, universal electrical power for all.

Tesla's radio frequency oscillators are still not fully understood. Tesla was a visionary, a mystic, and a genius. His vision was well over a century ahead of his contemporaries. In many respects it is still far ahead of the thinking in mainstream Western science. In working with radio waves Tesla created his most famous invention, the Tesla Coil. Invented in 1891,

83

the Tesla Coil is widely used in radio and television sets and other electronic equipment today.

Throughout this period Tesla gave the famous exhibitions in his laboratory. These demonstrations were in part conducted to allay fears of alternating current, and partly because he got a kick out of blowing people's minds. Do you remember Victor Frankenstein's lab in the Universal Pictures *Frankenstein* films of the 1930s? Those images were based on the stories handed down from those who had witnessed Tesla's demonstrations. Tesla wasn't a mad scientist, he was *the* Mad Scientist.

It was in his New York lab that Tesla experimented with vibration physics and the mechanical oscillator that nearly flattened New York City. Tesla's artificial earthquake was caused by a device he constructed in his lab to demonstrate the principle of harmonic resonance.

Continuing to quote from Jason Jeffrey's *New Dawn* article "Earthquakes: Natural or Man-Made?," Jeffrey wrote:

> In his Manhattan lab, Tesla built mechanical vibrators and tested their powers. One experiment got out of hand.
>
> Tesla attached a powerful little vibrator driven by compressed air to a steel pillar. Leaving it there, he went about his business. Meanwhile, down the street, a violent quaking built up, shaking down plaster, bursting plumbing, cracking windows, and breaking heavy machinery off its anchorages.
>
> Tesla's vibrator had found the resonant frequency of a deep sandy layer of subsoil beneath his building, setting off a small earthquake. Soon Tesla's own building began to quake. It is reported that just as the police broke into his lab, Tesla was seen smashing the device with a sledge hammer, the only way he could promptly stop it.
>
> In a similar experiment, on an evening walk through the city, Tesla attached a battery powered vibrator, described as being the size of an alarm clock, to the

steel framework of a building under construction. He adjusted it to a suitable frequency and set the structure into resonant vibration.

The structure shook, and so did the earth under his feet. Tesla later boasted he could shake down the Empire State Building with such a device. If this claim was not extravagant enough, he went on to say a large-scale resonant vibration was capable of splitting the Earth in half.

Over 40 years later, the *New York American* ran an article on 11 July 1935 entitled: "Tesla's Controlled Earthquakes." It stated that Tesla's "experiments in transmitting mechanical vibrations through the earth -- called by him 'the art of telegeodynamics' -- were roughly described by the scientist as a sort of 'controlled earthquake'." The article quoted Tesla as saying:

> The rhythmical vibrations pass through the earth with almost no loss of energy... it becomes possible to convey mechanical effects to the greatest terrestrial distances and produce all kinds of unique effects... The invention could be used with destructive effect in war...

In an interview in *The World Today*, February 1912, Tesla said that it would be possible to split the planet by combining vibrations with the correct resonance of the earth itself:

> Within a few weeks, I could set the earth's crust into such a state of vibrations that it would rise and fall hundreds of feet, throwing rivers out of their beds, wrecking buildings and practically destroying civilization.

Just before the turn of the century Tesla began his fateful experiments with wireless transmission of electrical power. He moved from New York to Colorado Springs, Colorado, where he built a new laboratory from the ground up to

85

there less than one year

develop his theories. The Colorado Springs lab contained the largest Tesla Coil ever built. He called it a "Magnifying Transmitter" (today such devices are call TMTs for Tesla Magnifying Transmitter). It was capable of generating some 300,000 watts of power.

The Patent Office often takes a decade or more to process an application. Tesla did not receive a patent on his magnifying transmitter until 1914. He applied for a patent to "Transmit Electrical Energy Through the Natural Mediums" in 1900, for which a patent was issued in 1905.

It was in Colorado Springs, where he stayed from May 1899 until early 1900, that Tesla made what he regarded as his most important discovery, and the one that led to his downfall — terrestrial stationary waves. Simply put, using this discovery, Tesla believed that energy could be pumped into the earth at any node of these waves, and could be extracted at any other. He demonstrated this on at least two occasions. In one experiment he lit 200 of Mr. Edison's biggest lamps, without wires, from a distance of 25 miles.

His lab pumped electricity into the earth using a 200 foot pole topped by a large copper sphere. With it he generated potentials that discharged lightning bolts up to 135 feet long. Thunder from the released energy could be heard in the town of Cripple Creek, some 15 miles away.

People walking along the streets in Colorado Springs were reported to have been amazed to see sparks jumping between their feet and the ground when Tesla's earth zapper was in operation. All over town flames of electricity shot from taps when unsuspecting citizens turned them on for a drink of water. Light bulbs within 100 feet of the experimental tower were reported to continue glowing long after they had been turned off. Several times during these experiments lightning storms were created, setting hundreds of fires across the state. Horses in at least one livery stable received shocks through their metal shoes and bolted from their stalls. Many other species were affected as well. Years later the local newspaper reported that during these experiments butterflies had became electrified and "helplessly swirled in circles - their wings spouting blue halos of 'St. Elmo's Fire.'"

At one point during the experiments in Colorado Springs he became convinced that he had received signals from another planet. Today it is believed that he had inadvertently discovered radio astronomy. The claim at the time, though, was met with derision in many scientific journals. That, along with his claims of having proven the workability of "broadcast power," and his increasingly strange behavior, earned him the dubious distinction of being widely regarded as the archetypical "mad scientist" by the nation's press.

SCALAR WEAPONS

Tesla's earthquake machine transmitted mechanical energy through the earth, and his TMTs may have done much the same with electricity. Is there a connection between these and Secretary Cohen's comments that "others" could "set off earthquakes, volcanoes remotely through the use of electromagnetic waves"?

Dr. Andrija Puharich, M.D., LL.D., (1918 – 1995) an experimental researcher and physician, was another of the world's leading "mad" scientists, with numerous patents granted in medicine and electronics. His primary work had been to bridge parapsychology and medicine. Some regard him as the "father" of the modern American "New Age" movement. Puharich was a supportive biographer of the spoon-bending Uri Geller and before that he had favorably investigated the Brazilian psychic surgeon Arigo. Puharich studied and designed devices emitting extremely low frequency (ELF) electromagnetic waves that he believed could affect the mind. In 1983 he was granted U.S. patent 4,394,230 for a "Method and Apparatus for Splitting Water Molecules." This method, *Wikipedia* tells us, "... would reportedly split water molecules into Hydrogen and Oxygen with a net energy gain, and is essentially a perpetual energy device that many believe violates the first law of thermodynamics."

Weather Warfare

In January 1978, Dr. Puharich issued a detailed research paper titled "Global Magnetic Warfare—A Layman's View of Certain Artificially Induced Unusual Effects On The Planet Earth during 1976 and 1977." He was primarily looking into the Soviet experiments with Tesla Magnifying Transmitters (TMT). Controlled earthquakes, he believed, were part and parcel of that work. Of them he wrote: "Of the many great earthquakes of 1976, there is one that demands special attention—the July 28, 1976 Tangshan, China earthquake."

That earthquake destroyed the city and killed at least 650,000 people. Nearly a year after the quake the *New York Times* reported on 5 June 1977 that "just before the first tremor at 3:42 AM, the sky lit up 'like daylight.' The multi-hued lights, mainly white and red, were seen up to 200 miles away. Leafs on many trees were burned to a crisp and growing vegetables were scorched on one side, as if by a fireball." Numerous researchers have since become convinced that a TMT was used to affect that quake.

In *FER DE LANCE: A Briefing On Soviet Scalar Electromagnetic Weapons*, Lt. Col Thomas E. Bearden (ret) (Ph.D., MS [nuclear engineering], BS [mathematics - minor electronic engineering] and co-inventor of the Motionless Electromagnetic Generator) contends that "the future of humanity has been hi-jacked for more than 50 years by the weaponization of scalar electromagnetics and … the Western scientific community has been blind-sided by dogmatic adherence to 1867 electrical theory." Elsewhere in *FER DE LANCE* he wrote:

> …The first KGB weather engineering tests over the U.S., using their relatively new interferometers, produced signatures of anomalous perfectly round holes appearing in clouds. These experiments started in 1967 or thereabouts. The Russians gave us the very severe "deep freeze" winter of 1967, as an initial weather-engineering test of their energetics interferometry weapons.

Lecturing at a symposium of the United States Psychotronics Association in 1981, Bearden defined their //energetics interferometry weapons as TMTs, saying:

> Tesla found that he could set up standing waves... in the earth (the molten core of the earth, or, just set it up through the rocks -- the telluric activity in the rocks would furnish activity into these waves and one would get more potential energy in those waves than he put in. He called the concept -- the magnifying transmitter.

Bearden elsewhere described the working of TMTs:

> They will go through anything. What you do is set up a standing wave through the Earth and the molten core of the Earth begins to feed that wave (We are talking Tesla now). When you have that standing wave, you have set up a triode. What you've done is that the molten core of the earth is feeding the energy and its like your signal--that you are putting in--is gating the gate of a triode....then what you do is that you change the frequency. If you change the frequency one way (dephase it) you dump the energy up in the atmosphere beyond the point on the other side of the earth that you focused upon. You start ionizing the air, you can change the weather flow patterns (Jet streams etc) — you can change all that — if you dump it gradually — real gradually — you influence the heck out of the weather...its a great weather machine. If you dump it sharply, you don't get a little ionization like that...you will get fireballs and flashes (Plasma & Earthquakes) that will come down on the surface of the earth...you can cause enormous weather changes over entire regions by playing that thing back and forth....

Could this have been what Secretary Cohen meant? If they can set off earthquakes, via electromagnetic waves or other means, what about tsunamis ("tidal waves")? Bearden claims that after the breakup of the Soviet Union, Russian scientists sold this technology to other hostile elements such as the *Yakusa*, the Japanese "Mafia," who used it to create the Asian Christmas tsunami!

PROJECT SEAL

Jeff Wells, a Canadian author and satirist who describes himself as "cautiously pessimistic," wrote in his blog:

> While I think a natural cause to the Sumatra tsunami remains the most likely explanation — Mother Nature gives great plausible deniability — the story of Project Seal is further proof that the military has a different take than many of us on what is unthinkable. And this was thinkable 60 years ago.
>
> Even among dissidents and the newly bestirred, there are many eyes squeezed shut to the ongoing presumption of the Pentagon to fold forces of the natural world into its mandate. I can understand why. It's scary as hell to doubt the natural provenance of such tremendous forces, and to suggest a human hand is to invite the harshest ridicule. How fortunate for the Pentagon.

As he said, "Mother Nature gives great plausible deniability." Therein lies the appeal of geophysical warfare, and not just to rogue states and terrorists. Covert operations have always been a part of "statecraft." Is it possible that competing nations have been secretly threatening each other with environmental mayhem as diplomatic bargaining chips? Could the Tangshan, China, earthquake have been retaliation for a diplomatic failure? What about the Sumatra tsunami,

Christmas 2004? Could it have been the ultimate drug deal gone bad?

Have those "ingenious minds" Cohen spoke of come up with a tsunami bomb? Unfortunately, yes.

Here are two stories by Eugene Bingham from *The New Zealand Herald* that tell the shocking tale of *Project Seal.* The 25 September 1999 edition of the paper carried the piece, "Tsunami Bomb NZ's Devastating War Secret," which revealed:

1945

> Top-secret wartime experiments were conducted off the coast of Auckland to perfect a tidal wave bomb, declassified files reveal.
>
> An Auckland University professor seconded to the Army set off a series of underwater explosions triggering mini-tidal waves at Whangaparaoa in 1944 and 1945.
>
> Professor Thomas Leech's work was considered so significant that United States defence chiefs said that if the project had been completed before the end of the war it could have played a role as effective as that of the atom bomb.
>
> Details of the tsunami bomb, known as Project Seal, are contained in 53-year-old documents released by the Ministry of Foreign Affairs and Trade.
>
> Papers stamped "top secret" show the US and British military were eager for Seal to be developed in the post-war years too. They even considered sending Professor Leech to Bikini Atoll to view the US nuclear tests and see if they had any application to his work.
>
> He did not make the visit, although a member of the US board of assessors of atomic tests, Dr Karl Compton, was sent to New Zealand.
>
> "Dr Compton is impressed with Professor Leech's deductions on the Seal project and is prepared to recommend to the Joint Chiefs of Staff that all technical data from the test relevant to the Seal project should

91

be made available to the New Zealand Government for further study by Professor Leech," said a July 1946 letter from Washington to Wellington.

Professor Leech, who died in his native Australia in 1973, was the university's dean of engineering from 1940 to 1950.

News of his being awarded a CBE in 1947 for research on a weapon led to speculation in newspapers around the world about what was being developed.

Though high-ranking New Zealand and US officers spoke out in support of the research, no details of it were released because the work was on-going.

Three days later Eugene Bingham and *The New Zealand Herald* elaborated on the above with "Devastating Tsunami Bomb Viable, Say Experts."

Tsunami experts believe a bomb secretly tested off the coast of Auckland 50 years ago could be developed to devastating effect.

University of Waikato researchers believe a modern approach to the wartime idea tested off Whangaparaoa could produce waves up to 30m high.

Dr Willem de Lange, of the Department of Earth Sciences, said studies proved that while a single explosion was not necessarily effective, a series of explosions could have a significant impact.

"It's a bit like sliding backwards and forwards in a bath - the waves grow higher," Dr de Lange said yesterday.

He was responding to a Weekend Herald report of experiments at Whangaparaoa in 1944-45 to create a tidal wave bomb. The top-secret work by the late Professor Tom Leech was detailed in 53-year-old papers released by the Ministry of Foreign Affairs and Trade.

Dr de Lange said a coastal marine group from

the university recently studied the likely impacts of underwater volcanic explosions.

Their work concluded that the next eruption in the Auckland region was likely to be under water, given the large amount of water around the city.

But tests showed a single explosion in the Hauraki Gulf would not trigger much of a tsunami.

"For most places the wave was less than 1m high, but it could be a bit more in the Tamaki Estuary."

Dr de Lange said the waves were not high because the energy was projected upwards, not sideways. He believed the same principle would be true for a tsunami bomb.

"You can't confine the energy. Once the explosion gets big enough, all of its energy goes into the atmosphere and not into the water. But one of the things we discovered was if you had a series of explosions in the same place, it's much more effective and can produce much bigger waves."

So it would seem that in the mid-1970s the United States military was aware of tests of bombs that had set off earthquakes in Nevada and tsunamis off the coast of New Zealand, and had participated in attempts to effect changes in climate and weather patterns over Southeast Asia — the very things that would become expressly prohibited by the EnMod Convention!

When Dr. MacDonald published his numerous papers and articles on future weapons technology — weapons that could exert control over the weather and climate and trigger earthquakes — he was a senior official at the top of the military's research and development "food chain." It looks to me like his "speculations" were based on insider knowledge of what technologies were actually under development. I believe he was trying to give us an advance warning of where he knew the military-industrial-academic complex was taking us. *Are we there yet?*

Weather Warfare

After conducting his investigations Senator Pell clearly thought we were there, or would be soon and led the world in demanding such weapons be banned. As we shall see in the next chapter there is indeed a United Nations treaty making the hostile use of environmental modification a crime against humanity — but is it enforceable? Can it actually protect us?

Table 1

Operational Capabilities Matrix

DEGRADE ENEMY FORCES	ENHANCE FRIENDLY FORCES
Precipitation Enhancement	**Precipitation Avoidance**
- Flood Lines of Communication - Reduce PGM/Recce Effectiveness - Decrease Comfort Level/Morale	- Maintain/Improve LOC - Maintain Visibility - Maintain Comfort Level/Morale
Storm Enhancement	**Storm Modification**
- Deny Operations	- Choose Battlespace Environment
Precipitation Denial	**Space Weather**
- Deny Fresh Water - Induce Drought	- Improve Communication Reliability - Intercept Enemy Transmissions - Revitalize Space Assets
Space Weather	
- Disrupt Communications/Radar - Disable/Destroy Space Assets	**Fog and Cloud Generation**
	- Increase Concealment
Fog and Cloud Removal	**Fog and Cloud Removal**
- Deny Concealment - Increase Vulnerability to PGM/Recce	- Maintain Airfield Operations - Enhance PGM Effectiveness
Detect Hostile Weather Activities	**Defend against Enemy Capabilities**

From *Weather as a Force Multiplier: Owning the Weather in 2025,* an Air Force Study in 1996.

94

CHAPTER FOUR

ENMOD LEGISLATION

> *If man can modify the weather, he will obviously modify it for military purposes. It is no coincidence that the U.S. Army, Navy, Air Force and Signal Corps have been deeply involved in weather modification research and development. Weather is a weapon, and the general who has control over the weather is in control of an opponent less well armed... The idea of clobbering an enemy with a blizzard, or starving him with an artificial drought still sounds like science fiction. But so did talk of atom bombs before 1945.*

So wrote author Daniel S. Halacy Jr., in his book, *The Weather Changers*, published in 1968.

In the 1950s weather control and environmental modification were the stuff of science fiction and popular "frontiers of science" stories—such as "Weather Made To Order?" by the previously mentioned Capt. H. T. Orville that appeared on the cover of *Collier's* in May of 1954. At the time Capt. Orville was an advisor to President Dwight D. Eisenhower on matters of weather modification.

In 1967 the U.S. Senate passed the Magnusson Bill that authorized the Secretary of Commerce to accelerate programs of applied research, development and experimentation in weather and climate modification (EnMod). The bill allocated $12 million, $30 million and $40 million over the next three years, respectively. They projected expenditures of some $149 million annually by 1970.

It can be argued that by the beginning of the 1970s portions of the U.S. government and/or military viewed EnMod research as having transitioned from the "basic research" stage to the "operational" stage. Experiments were occurring — or had occurred — in 22 countries, including Argentina, Australia, Canada, Iran, Israel, Kenya, Italy, France, South Africa, Congo and the U.S.S.R. Airborne seeding programs were undertaken to combat drought in the Philippines, Okinawa, Africa and Texas. Fog clearing had become a standard operation at airports, as had hailstorm abatement, which had been proven successful in several parts of the world. Forest fire control had been carried out in Alaska and watershed seeding was widely practiced, while lake storm snow redistribution was under extensive investigation. By FY1973 there were over 700 degreed scientists and engineers in the U.S. whose major occupation was weather modification.

And then it all changed. In 1978 The United States became a signatory to the United Nations Convention on the Prohibition of Military or Any Other Hostile Use of Environmental Modification Techniques (EnMod Convention or ENMOD for short). The EnMod Convention prohibits the use of techniques that would have widespread, long-lasting or severe effects through deliberate manipulation of natural processes and cause such phenomena as earthquakes, tidal waves and changes in climate and weather patterns.

Returning to Keith Harmon Snow's massive report "Out of the Blue: Black Programs, Space Drones & The Unveiling of U.S. Military Offensives in Weather as a Weapon;" he tells us:

> In 1976, U.S. government officials outlined 50 experimental projects and 20 actual pilot programs costing upwards of $100 million over the next eight years.
>
> It was an explosive subject, up [through] the 1970s but, after 1977, ENMOD interest seemed to disappear almost overnight. In other words, after decades of intense research and development, after billions of dollars of investment, after major institutions and

CARTER ELECTED M/1976 PRES 1977-1980

governmental bodies were created and charged with oversight of ENMOD and its many peripheral issues, and after the entire reorganization of the U.S. Government to channel and guide and map out the future of this new and promising military and civilian "technology" — said to be more important than the atom bomb — everything stopped.

Or did it?

It was as if a huge curtain fell over the subject as all research, all institutional interests, huge salaries and thousands of jobs — vanished. And the mass media stopped reporting anything and everything as if struck by plague. That — sudden and total silence — is perhaps the most telling and suspicious indication of the secrecy and denial that the ENMOD arena was shackled with. Today it is almost as if it never happened.

Could it be that the US government said, "Oh gee, we can't do that any more" and just gave up on military EnMod — or did the whole program go "black"?

By the late 1970s weather control and environmental modification had become not only a scientific fact, but the subject of national and international law. Many researchers have commented that they would not have bothered to create a law prohibiting one nation from attacking another with earthquakes, tidal waves and devastating weather unless they actually had, or thought they would soon have, the technology to do it. It is my contention that all the ENMOD treaty did was force this line of research, and use, into hiding.

THE NATIONAL WEATHER MODIFICATION POLICY ACT OF 1976

One of the first US legislative steps toward ratifying the EnMod Convention was the *National Weather Modification Policy Act of 1976* a follow up on the Magnusson Bill. It directed the Secretary of Commerce to conduct a comprehensive study of the status of weather modification science and technology and to submit to the President and the Congress a report on the findings, conclusions, and recommendations of the study. To conduct that study the Secretary of Commerce created the Weather Modification Advisory Board (WMAB). That Board gave its findings, via the Secretary of Commerce, in *The National Weather Modification Policies and Programs* report of 1979, which was submitted to Congress in November of that year.

The Weather Modification Advisory Board (WMAB) proposed the establishment of a comprehensive, and well funded, continuing research program by the federal government into weather modification. It also proposed multilateral research and development agreements and comprehensive international accords, such as the ENMOD Convention that would soon follow. The WMAB concluded, "The prime requirement of a national weather modification policy is to learn more about the atmosphere itself" — *D'oh!*

The Secretary of Commerce was given the task "to conduct a comprehensive study of the status of weather modification science and technology" and submit a report on his findings because one of the agencies of his Department was (and still is) the National Oceanic and Atmospheric Administration (NOAA), which was the major federal civilian player in weather modification at the time. I always found it amusing that the name of the agency given the task of understanding rain and oceans was pronounced "Noah."

NOAA got a lot of good press in the '60s and '70s for its many research projects. For example, NOAA has pursued an experimental program since 1967 to study the potential for augmenting rainfall from subtropical cumulus clouds, the results of which have had important implications for developing countries in the tropics, where clouds of this type account for most of the rainfall.

At the time of its 1979 report the Weather Modification Advisory Board (WMAB) indicated that "NOAA regards itself and is regarded by Congress, as the focal Agency for

98

matters having to do with the atmosphere and the oceans."
At that time NOAA had several research laboratories under
its domain:

- NOAA's National Severe Storms Laboratory in
Norman, Oklahoma
- NOAA's National Hurricane and Experimental
Meteorology Laboratory and NOAA's Research Facilities
Center, both in Miami, Florida
- NOAA's Atmospheric Physics and Chemistry
Laboratory and NOAA's Wave Propagation Laboratory,
both in Boulder, Colorado
- NOAA's Geophysical Fluid Dynamics Laboratory in
Princeton, New Jersey, and
- NOAA's Air Resources Laboratories which were
located at several sites around the country.
- NOAA was also responsible for the federal
government's civilian atmospheric services, including:
The National Weather Service
The National Environmental Satellite Service and
The Environmental Data and Information Service.

As the WMAB said of NOAA in its report: "It is thus the
primary source within the Federal government of the talents,
observations, data, and information essential to weather
modification field programs and evaluations."

Several other bills related to weather and environmental
modification have been introduced into the US Congress since
the *National Weather Modification Policy Act of 1976*. Before we
get to the EnMod Convention I think looking at a couple of
these might prove enlightening.

SPACE PRESERVATION ACT OF 2001

Chemtrails, radio frequency weapons and mind control
are all mentioned by name in House Bill HR 2977, the Space
Preservation Act of 2001, which was introduced into the 107th
Congress, 1st Session by Congressman Dennis Kucinich of
Ohio, who later vied for the Democratic Party Presidential

nomination in run up to the 2004 election.

Kucinich was elected to congress in 1996 and has since served on and occasionally chaired several committees, including the Armed Services Oversight Committee, and the Committee on Government Reform, Subcommittee on National Security, Emerging Threats and International Relations. Additionally Kucinich authored and/or co-sponsored legislation to create a national health care system, lower the costs of prescription drugs, provide economic development through infrastructure improvements, abolish the death penalty, provide universal pre-kindergarten to all 3- 4- and 5-year-olds, create a Department of Peace, regulate genetically engineered foods, repeal the USA PATRIOT Act, and provide tax relief to working class families. Kucinich is proud of having been honored by Public Citizen, the Sierra Club, Friends of the Earth and the League of Conservation Voters as a champion for clean air, clean water and an unspoiled earth. Kucinich has twice been an official United States delegate to the United Nations Convention on Climate Change (1998, 2004) and attended the 2002 World Summit on Sustainable Development in Johannesburg, South Africa.

The purpose of the Space Preservation Act of 2001 was:

> To preserve the cooperative, peaceful uses of space for the benefit of all humankind by permanently prohibiting the basing of weapons in space by the United States, and to require the President to take action to adopt and implement a world treaty banning space-based weapons.

Section 2 of the bill specifically reaffirmed that it "is the policy of the United States that activities in space should be devoted to peaceful purposes for the benefit of all mankind," as expressed in section 102(a) of the National Aeronautics and Space Act of 1958 (42 U.S.C. 2451(a)).

Sections 3 and 4 attempted to place a permanent ban on the basing of weapons in space. These called for the President

100

to:

    (1) implement a permanent ban on space-based weapons of the United States and remove from space any existing space-based weapons of the United States; and

    (2) immediately order the permanent termination of research and development, testing, manufacturing, production, and deployment of all space-based weapons of the United States and their components.

And ordered that:

    The President shall direct the United States representatives to the United Nations and other international organizations to immediately work toward negotiating, adopting, and implementing a world agreement banning space-based weapons.

The bill also required the President to submit to Congress a report on the implementation of the permanent ban on space-based weapons required by these sections and on the progress being made toward negotiating, adopting, and implementing a world–wide agreement banning space-based weapons — within 90 days of the Act being enacted and every 90 days thereafter, *forever!*

The bill also expressly stipulated that nothing in the Act should be construed as prohibiting the use of funds for space exploration; space research and development; or testing, manufacturing, or production that is not related to space-based weapons or systems; or civil, commercial, or defense activities that are not related to space-based weapons or systems.

The Act defined `space' as meaning all space extending upward from an altitude greater than 60 kilometers above the surface of the earth and any celestial body in such space. And defined the terms `weapon' and `weapons system' to mean a device capable of any of the following:

(i) Damaging or destroying an object (whether in outer space, in the atmosphere, or on earth) by--

(I) firing one or more projectiles to collide with that object;

(II) detonating one or more explosive devices in close proximity to that object;

✓ (III) directing a source of energy (including molecular or atomic energy, subatomic particle beams, electromagnetic radiation, plasma, or extremely low frequency (ELF) or ultra low frequency (ULF) energy radiation) against that object; or

(IV) any other unacknowledged or as yet undeveloped means.

(ii) Inflicting death or injury on, or damaging or destroying, a person (or the biological life, bodily health, mental health, or physical and economic well-being of a person) —

(I) through the use of any of the means described in clause (i) or subparagraph (B);

(II) through the use of land-based, sea-based, or space-based systems using radiation, electromagnetic, psychotronic, sonic, laser, or other energies directed at individual persons or targeted populations for the purpose of information war, mood management, or mind control of such persons or populations; or

(III) by expelling chemical or biological agents in the vicinity of a person.

(B) Such terms include exotic weapons systems such as —

→ (i) electronic, psychotronic, or information weapons;

(ii) chemtrails;

(iii) high altitude ultra low frequency weapons systems;

(iv) plasma, electromagnetic, sonic, or ultrasonic

102

weapons;

(v) laser weapons systems;

(vi) strategic, theater, tactical, or extraterrestrial weapons; and

(vii) chemical, biological, environmental, climate, or tectonic weapons.

(C) The term `exotic weapons systems' includes weapons designed to damage space or natural ecosystems (such as the ionosphere and upper atmosphere) or climate, weather, and tectonic systems with the purpose of inducing damage or destruction upon a target population or region on earth or in space.

Note that "(iii) high altitude ultra low frequency weapons systems" sure sounds like HAARP!

Many people believe that this bill, the first official US Government document to contain the term "chemtrail," is an official acknowledgment of the possibility, at least, of the existence of such weapons systems.

Bob Fitrakis, of *Columbus Alive*, reported in January of 2002 that:

> Alive asked Kucinich why he would introduce a bill banning so-called chemtrails when the U.S. government routinely denies such things exist and the U.S. Air Force has routinely called chemtrail sightings "a hoax."
>
> "The truth is there's an entire program in the Department of Defense, 'Vision for 2020,' that's developing these weapons," Kucinich responded.

Much cooler heads apparently prevailed, as all references to Chemtrails, mind control and other such controversial terms were omitted from the versions of the bill re-introduced by Kucinich in 2002 as HR 3616 and in 2003 as HR 3657.

WEATHER MODIFICATION RESEARCH AND TECHNOLOGY TRANSFER AUTHORIZATION ACT OF 2005

Senator Kay Bailey Hutchison of Texas introduced the *Weather Modification Research and Technology Transfer Authorization Act of 2005* into the U.S. Senate on 3 March 2005 as S. 517. A similar bill was introduced into the House as H.R. 2995. Its purpose is to create an updated WMAB by establishing the *Weather Modification Operations and Research Board*.

The Senate Commerce Committee's Science and Space Subcommittee and its Disaster Prevention and Prediction Subcommittee held a joint hearing on the bill on 10 November 2005. Presenting evidence in favor of the bill were Dr. Tom DeFelice, Past President, Weather Modification Association; Dr. Joe Golden, Senior Research Scientist, University of Colorado's Cooperative Institute for Research in the Environmental Sciences; and Dr. Michael Garstang, Chair of the Committee on Critical Issues in Weather Modification Research, National Research Council of The National Academies.

On 8 December 2005 it was voted out of the Committee on Commerce, Science, and Transportation and placed on the Senate Legislative Calendar under General Orders, Calendar Number 319.

If passed, which seems likely, this bill:

Directs the Director of the Office of Science and Technology Policy to establish a Weather Modification Subcommittee to coordinate a national research program on weather modification. Requires the Subcommittee to include representatives from: (1) the National Oceanic and Atmospheric Administration (NOAA); (2) the National Science Foundation (NSF); and (3) the National Aeronautics and Space Administration (NASA). Provides for a representative from NOAA and a representative from NSF to serve together as co-chairs of such Subcommittee.

Requires the Director to develop and submit a plan for coordinated federal activities under the program, which shall: (1) for a ten-year period, establish the goals and priorities for federal research that most effectively advances scientific understanding of weather modification; (2) describe specific activities required to achieve such goals and priorities, including funding of competitive research grants, training and support for scientists, and participation in international research efforts; (3) identify and address, as appropriate, relevant programs and activities of the federal agencies and departments that would contribute to the program; (4) consider and use, as appropriate, reports and studies conducted by federal agencies and departments, and other expert scientific bodies, including the National Research Council report on Critical Issues in Weather Modification Research; (5) make recommendations for the coordination of program activities with weather modification activities of other national and international organizations; (6) incorporate recommendations from the Weather Modification Research Advisory Board; and (7) estimate federal funding for research activities to be conducted under the program.

Specifies activities related to weather modification that may be included under the program, including: (1) interdisciplinary research and coordination of research and activities to improve understanding of processes relating to weather modification, including cloud modeling, cloud seeding, improving forecast and decision-making technologies, related severe weather research, and potential adverse affects of weather modification; (2) development, through partnerships among federal agencies, states, and academic institutions, of new technologies and approaches

105

for weather modification; and (3) scholarships and educational opportunities that encourage an interdisciplinary approach to weather modification.

Requires the Director to prepare and submit to the President and Congress annual reports on the activities conducted pursuant to this Act respecting the Weather Modification Subcommittee, including: (1) a summary of the achievements of federal weather modification research; (2) an analysis of the progress made toward achieving the goals and objectives of the plan; (3) a copy or summary of the plan and any changes made to it; (4) a summary of agency budgets for weather modification activities; (5) any recommendations regarding additional action or legislation that may be required to assist in achieving the purposes of this Act; (6) a description of the relationship between research conducted on weather modification and research conducted pursuant to the Global Change Research Act of 1990, as well as research on weather forecasting and prediction; and (7) a description of any potential adverse consequences on life, property, or water resource availability from weather modification efforts, and any suggested means of mitigating or reducing such consequences if such efforts are undertaken.

(Sec. 5) Establishes in the Office of Science and Technology Policy the Weather Modification Research Advisory Board to: (1) make recommendations to the Weather Modification Subcommittee on matters related to weather modification; and (2) advise such Subcommittee on the research and development, studies, and investigations with respect to potential uses of technologies and observation systems for weather modification research and assessments and evaluations of the efficacy of weather modification, both purposeful, (including cloud-seeding operations) and inadvertent (including downwind effects and anthropogenic effects).

(Sec. 6) Instructs U.S. departments and agencies and any other public or private agencies and institutions that receive research funds from the United States related to weather modification to give full support and cooperation to the Weather Modification Subcommittee.

A surprisingly large number of people have expressed opposition to this legislation. Some, ignorant of how far weather modification has progressed, are shocked to hear that the U.S. is considering anything dealing with intentionally altering the weather. Others, more astute, have noticed that this proposed Weather Modification Operations and Research Board would have no oversight by those affected by these operations.

As drafted this Board would only consist of scientists representing academic organizations and governmental agencies already involved in weather modification, with no input from farmers, business owners, ranchers, environmentalists, etc. Some see this as a case of foxes requesting to be put in charge of henhouse operations. If there were some sort of conspiracy this would be a very neat way of ensuring that only the science the conspirators wanted done would get funded, and only those reports they wanted the public to see would ever reach print. But that's crazy talk, right?

THE CONVENTION ON THE PROHIBITION OF MILITARY OR ANY OTHER HOSTILE USE OF ENVIRONMENTAL MODIFICATION TECHNIQUES (ENMOD)

Perhaps the United States' first steps in international weather accords were taken by the 90th Congress, which in 1968 passed Concurrent Resolution 67. That Resolution declared that the official policy of the United States was to cooperate with other nations in the weather modification field.

A formal agreement on weather modification information exchange was signed between the United States and Canada in March of 1975. That agreement was inked after emotional concern surfaced over a US proposal to engage in commercial cloud seeding in northern Washington State, near the Canadian border. In that agreement the USA and Canada promised to provide advance notification and consultation with respect to activities conducted within 200 miles of the international boundary, or whenever either party believed the effects of weather modification activities would be significant to the other party. Similarly, the United States initiated negotiations with Mexico in 1978 toward the possibility of a joint experimental program on hurricanes in the Eastern Pacific. Technical discussions with a number of other countries have taken place over the years since.

After the revelations of Operation Popeye the die was cast. Senator Pell, with Representatives Gilbert Gude of Maryland and the above-mentioned Donald Fraser, became the three leading legislative critics of American military research into weather and environmental modification. Together, they sent a letter to President Gerald Ford urging increased government support for the peaceful uses of such modification. They also urged that all such research and operations, military and non-military, be overseen by a civilian agency answerable to Congress and the President. Their recommendations were mostly ignored.

Another influential voice calling for limits on the military's ability to wreck the environment was Lowell Ponte. He had worked as a DoD consultant on environmental and bizarre weapons for the International Research & Technology Corporation of Washington, D.C. and, later, became editor of *Skeptic Magazine*. In his book *The Cooling* he described the Congressional hearings mounted by Pell, Fraser and Gude thusly: "What emerged was an awesome picture of far-ranging research and experimentation by the Department of Defense into ways environmental tampering could be used as a weapon." Among the unthinkable things the military

108

considered were investigations into whether lasers and chemicals could create an artificial hole in the ozone layer over an enemy, causing damage to crops and human health through exposure to the sun's ultraviolet rays. This revelation led to that technology being explicitly banned by the ENMOD Convention.

Senator Pell did a lot of stumping and article writing to force the world to act. In one article he wrote:

> Apart from the sheer horror of the prospect of unbridled environmental warfare, there is, I believe, another compelling reason to ban such action. We know, or should know, by now, that no nation can maintain for long a monopoly on new warfare technology. If we can develop weather warfare techniques, so can and will other major powers. Experience has taught us that the weapons that make us feel secure today, will make us feel very insecure indeed when our adversaries possess the same capabilities.

In *The Cooling*, Lowell Ponte describes the events that led to the ENMOD Convention:

> During a summit meeting between President Nixon and Soviet Premier Leonid Brezhnev on July 3, 1974, the nations agreed to conduct discussions toward a ban on environmental warfare. Before the first of these discussions, set for Moscow in November, got underway, the Soviet Union introduced a resolution before the United Nations General Assembly to ban environmental warfare. When revised, the resolution was passed by the body 102 votes to none. The United States and half a dozen other nations abstained from the vote. Senator Pell suspected that the president felt miffed by the surprise Soviet action, a move that made it appear that the Soviet Union and not the United States had taken the lead in trying to ban environmental

modification. In fact, the Soviet resolution was similar to one passed by the North Atlantic Assembly in November 1972 and to another authored by Senator Pell and passed by an 82 to 10 vote by the United States Senate in July 1973.

Discussion between U.S. and Soviet negotiators resumed in Washington, D.C., on February 24, 1975. On August 21, 1975, the two nations presented their jointly produced draft treaty banning environmental modification as a weapon of war to the thirty-one-nation Geneva Disarmament Conference.

The EnMod Convention (ENMOD) was later passed by the United Nations General Assembly and opened for signature in 1977. It came into effect 5 October 1978, when it was certified by the required total of 20 nations.

ENMOD is often described as a "non-use" agreement because it prohibits the use of environmental modification as a weapon, rather than restricting the development of modification techniques. Many other arms control agreements take quite different approaches; for example, the Biological and Toxin Weapons Convention prohibits the development and stockpiling of a whole class of weapons, whereas most of the nuclear arms control agreements place restrictions on the number and/or types of weapons that may be deployed.

ENMOD's ratification by the United States, however, was hardly a done deal. Environmentalists, led by the Natural Resources Defense Council, the Sierra Club, the Environmental Policy Center, the Wilderness Society and the Federation of American Scientists, were intensely disappointed by the final text of ENMOD—so much so that they opposed US ratification.

Among their fears were the perception that the language defining modification techniques was so vague and the threshold for violations so high that ENMOD might legitimize the use of some environmental weapons. These fears seem to have been validated by the Air Force Air University white paper *Weather as a Force Multiplier: Owning the Weather in 2025.* As I will detail shortly, this paper lays out a surprising number of possible "interventions" that could modify the environment to enhance friendly forces or degrade enemy forces without violating the letter of ENMOD.

The environmentalists also noted that ENMOD's umbrella of protection was far too small, for it excluded weapons development and testing. Many researchers worldwide are convinced that weather weapons have been and are being developed and tested as you read this. There were also concerns that ENMOD's mechanism for verification is too weak, which also seems to be the case, as we shall see.

Despite a court order requiring that an environmental impact assessment of ENMOD be made prior to its submission to the Senate for ratification, and calls by a few environmentalists for a complete renegotiation of ENMOD, it soon became clear that world opinion strongly favored ratification. The environmental coalition opposing ENMOD withdrew their objections and changed their tactics to advocating improving the treaty through a rapid series of conferences of State Parties (signatories) to strengthen ENMOD's weaker provisions. Not surprisingly, the proposed conferences never came about.

As we will examine shortly, the environmentalist objections to ENMOD were dead-on. It is a treaty without teeth, and one that is now virtually ignored by the international community.

The clearest and most detailed critique of ENMOD is "Addressing Environmental Modification in Post-Cold War Conflict—The Convention on the Prohibition of Military or Any Other Hostile Use of Environmental Modification Techniques (ENMOD) and Related Agreements" by Susana Pimiento Chamorro and Edward Hammond of The Sunshine Project. This brilliant paper was written for presentation to the Civil Society Conference to Review ENMOD and Related Agreements on Hostile Modification of the Environment, which was held in Amsterdam in May 2001 and sponsored by the Sunshine Project, the Edmonds Institute, the Third

111

World Network, and the Transnational Institute. They began that paper with:

> Since the end of the Cold War, superpower confrontations have been replaced by new kinds of conflicts, tactics and weaponry. Military and political strategists have labeled narcotics, ethnic disputes (including indigenous peoples' grievances), competition for natural resources (including water), and protests against globalization as new or emerging threats to national and global security. Traditional "army versus army" confrontations have been supplanted by complex combinations of law enforcement, peacekeeping, economic and military measures. New kinds of technology, including biotechnology, have been adopted by militaries seeking to adapt to the changing face of conflict.
>
> Environmental treaties offer one avenue for constraint on environmental modification in times of conflict. Unfortunately, in recent years, it has proven difficult to gain general ratification of new environmental agreements. The US in particular has balked at ratifying several important agreements ranging from the Biodiversity Convention to the Kyoto Protocol. This in mind, the use of existing legal instruments would appear to be a more practical approach to constraint of environmental destruction than the creation of new treaties.
>
> Addressing emergent conflicts with an existing international legal instrument, however, may prove challenging. Many arms control treaties were tailored to suit the Cold War, not the 21st century. The Convention on the Prohibition of Military or Any Other Hostile Use of Environmental Modification Techniques (ENMOD; see text in Appendix) was intended to address technologies capable of modifying the environment for hostile purposes. Provocation of earthquakes, cloud

seeding, and manipulation of the atmosphere are examples of the kinds of possible attacks the authors of ENMOD foresaw. The treaty's preventative approach is both visionary and remarkable, considering that, at the time of negotiations, the technology to achieve many of the weapons contemplated was in its initial stages or entirely nonexistent.

Activities that could violate ENMOD include:
Triggering earthquakes
Manipulating ozone levels
Alteration of the ionosphere
Deforestation
Provoking flood or drought
Use of herbicides
Setting fires
Seeding clouds
Introduction of invasive species
Eradication of species
Creation of storms
Manipulation of El Niño / La Niña
Destruction of crops

Today, almost a quarter century after ENMOD was negotiated, the treaty is all but forgotten. Only 70 countries ratified the agreement. Twenty were required to bring it into force. Parties to ENMOD have met only twice, in 1982 and 1992. Despite the fact that, early on, ENMOD was recognized as flawed, no international civil society group has seriously tried to fix it.

Before we get into dissecting the failure of the EnMod Convention let's take a quick look at the history of international conventions limiting what belligerents can do in warfare, of which the EnMod Convention is an outgrowth.

THE LAW OF WAR

Although it may appear to be an incongruous

113

concept, the nations of the world have recognized the need to impose restrictions on the waging of war. War will necessarily result in death and injury to humans and the destruction of property; however, in the eyes of the international community, it need not be an unlimited exercise in cruelty and ruthlessness. The necessities of war must be conciliated with the laws of humanity. The resulting restrictions are regarded as the international Law Of Armed Conflict (LOAC), or the law of war.

These concepts... can be found throughout man's history. The ancient Hindu laws of Manu prohibited the use of barbed arrows because they exacerbated the injury upon their removal. The Romans considered the use of poisoned weapons to be unlawful. During the Middle Ages, the Pope condemned the crossbow, noting the appalling injuries it caused...

So wrote Major Joseph W. Cook, III; Major David P. Fiely; and Major Maura T. McGowan in their U.S. Air Force sponsored research paper, "Nonlethal Weapons: Technologies, Legalities, and Potential Policies."

In 1868, the Russian government issued an invitation to the International Military Commission "to examine the expediency of forbidding the use of certain projectiles in time of war between civilized nations." A new type of bullet was the cause of the Russian concern. These new "light explosives" or "inflammable projectiles," when used against human beings, were no more effective than an ordinary rifle bullet; however, they caused greater wounds and thus greatly aggravated the sufferings of the victim.

The resulting document, The Declaration of St. Petersburg, prohibited the use of explosive projectiles under 400 grams of weight. It was the first international treaty imposing restrictions on the conduct of war in modern times. The Declaration of St. Petersburg developed a line of reasoning governing the legality of weapons, which is found in its

Preamble:

Considering the progress of civilization should have the effect of alleviating as much as possible the calamities of war. The only legitimate object which states should endeavor to accomplish during war is to weaken the military forces of the enemy. It is sufficient to disable the greatest possible number of men, and this object would be exceeded by the employment of arms which uselessly aggravate the sufferings of disabled men or render their death inevitable. The use of such weapons would therefore, be contrary to the laws of humanity.

While most cultures around this globe have seen the need to restrain the horrors of war, it was not until the nineteenth century that these sentiments were codified into international law. The Declaration of St. Petersburg was followed in 29 July 1899 with the Hague Conventions, which codified the "laws and customs of war on land." These were amended several times during the following decade.

The Geneva conventions of 1929 and 1949 focused on ameliorating the conditions of civilians, prisoners of war, and the sick and wounded. These can be found in Treaties and Other International Acts Series (TIAS) (Washington, D.C.: US Department of State, 1956):

- Geneva Convention for the Amelioration of the Condition of the Wounded and Sick in the Armed Forces in the Field, 12 August 1949, TIAS 3362
- Geneva Convention for the Amelioration of the Condition of the Wounded and Sick and Shipwrecked Members, 12 August 1949, TIAS 3363
- Geneva Convention Relative to the Treatment of Prisoners of War, 12 August 1949, TIAS 3363
- Geneva Convention Relative to the Treatment of Civilian Persons in Time of War, 12 August 1949, TIAS 3365.

115

Additionally, a number of international treaties address the legitimacy of specific weapons. These include the following agreements:

• The 1925 Geneva Protocol for the Prohibition of the Use of Asphyxiating, Poisonous, or Other Gases and of Bacteriological Methods of Warfare, 17 June 1925

• Convention on the Prohibition of the Development, Production, and Stockpiling of Bacteriological (Biological) and Toxin Weapons and on Their Destruction, 10 April 1972

• Convention on the Prohibition of Military or Any Other Hostile Use of Environmental Modification Techniques, 10 December 1976

• Convention on Prohibitions or Restrictions on the Use of Certain Conventional Weapons Which May Be Deemed to Be Excessively Injurious or to Have Indiscriminate Effects, 10 October 1980,

• Protocol on Non Detectable Fragments, 10 October 1980

• Protocol on Prohibitions or Restrictions on the Use of Mines, BoobyTraps, and Other Devices, 10 October 1980.

• Convention on the Prohibition of the Use, Stockpiling, Production and Transfer of Anti-Personnel Mines and on Their Destruction, 1 March 1999.

In addition to the international EnMod Convention there have been other regional efforts to impose limits on environmental damage from war making. These have included the Central American Water Tribunal, which imposes moral sanctions in cases of severe pollution, and the Lusaka Agreement on Cooperative Enforcement Operations Directed at Illegal Trade in Wild Fauna and Flora, established by six African countries in 1999 to fight wildlife crime. The European Union established criminal responsibility for ecocide in 1988 with its Convention on the Protection of the Environment through Criminal Law.

116

LIMITATIONS ON A STATE'S RIGHTS IN CONFLICTS

*The right of belligerents to adopt means of
injuring the enemy is not unlimited.*
~Article 22. The Hague Regulations Respecting the
Laws and Customs of War on Land, 1907

International law recognizes the right of states to wage war based on the principles of sovereignty and self-defense. The right of states to engage in armed conflict, however, is not absolute. As we have seen above the law of war now places restrictions on the primary aspects of armed conflict, including:

- the definition of war
- relations between neutral and belligerent states
- and the conduct of war: that is, treatment of prisoners, wounded, civilians in occupied territories, enemy nationals and their property, non-military ships, and the types of weapons that may or may not be developed or deployed.

International rules on the conduct of war are intended to avoid unnecessary suffering or damage to combatants, civilian populations and property. But under the law of war, the definition of unnecessary is decidedly limited. Generally speaking, by declaring a military necessity, states can exempt themselves from the restrictions of the law of war and sidestep the restrictions it imposes. There are limitations, however, on states' ability to claim exceptions on the basis of military necessity.

In the case of *United States v. List*, the international military tribunal in Nuremberg following World War II determined that:

Military necessity permits a belligerent, subject to

the laws of war, to apply any amount and kind of force to compel the complete submission of the enemy with the least possible expenditure of time, life, and money... There must be some reasonable connection between the destruction of property and the overcoming of the enemy.

This means that the rules of international law must be followed even if it results in the loss of an advantage. *Kriegsraison*, the German doctrine of military necessity, was the belief that the ends justified the means — that a matter of "urgent necessity" could override the law of war. This principle was rejected in *United States v. Krupp*, when the Nuremberg tribunal held that:

...to claim that the law of war can be wantonly and at the sole discretion of any one belligerent be disregarded when he considered his own situation to be critical means nothing more than to abrogate the laws and customs of war entirely.

The Kellogg-Briand Pact of 1928, the so-called "Paris Peace Pact," was a treaty between the United States and other Powers concluded as a valiant effort to prevent World War II from occurring by having the nations who signed it formally renounce war as an instrument of national policy. Quick history: signed at Paris, 27 August 1928; ratification advised by the U.S. Senate, 16 January 1929; ratified by the President, 17 January 1929; instruments of ratification deposited at Washington by the United States of America, Australia, Dominion of Canada, Czechoslovakia, Germany, Great Britain, India, Irish Free State, Italy, New Zealand, and Union of South Africa, 2 March 1929: By Poland, 26 March 1929; by Belgium, 27 March 1929; by France, 22 April 1929; by Japan, 24 July 1929; proclaimed, 24 July 1929.

Article 1 of The Kellogg-Briand Pact provided:

The High Contracting Parties solemnly declare in the names of their respective people that they condemn recourse to war for the solutions of international controversies, and renounce it as an instrument of national policy in their relations with one another.

When signing the Pact Germany entered a reservation to the effect that it reserved the right to go to war in self-defense as determined by itself. Dr. Francis A. Boyle is a full professor with the University of Illinois College of Law in Champaign. A scholar in the areas of international law and human rights, Professor Boyle received a J.D. degree *magna cum laude* and A.M. and Ph.D. degrees in political science from Harvard University. Prior to joining the faculty at the College of Law, he was a teaching fellow at Harvard and an associate at its Center for International Affairs. Twenty-plus years of anti-nuclear advocacy may have earned him the world's record for number of anti-nuclear acquittals. In his book *The Criminality of Nuclear Deterrence* he wrote:

So when in 1945 the Nazi war criminals were prosecuted for crimes against peace on the basis of the Kellogg-Briand Pact, they basically argued that the Second World War was a war of self-defense as determined by the Nazi government, and therefore that the Nuremberg Tribunal had no competence to determine otherwise because of Germany's self-judging reservation. Needless to say, the Tribunal summarily rejected this preposterous argument and later convicted and sentenced to death several Nazi war criminals for the commission of crimes against peace, among other international crimes.

Article 6(a) of the 1945 Nuremberg Charter defines "crimes against peace" as follows:

(a) CRIMES AGAINST PEACE: namely,

119

> planning preparation, initiation or waging of a war
> of aggression, or a war in violation of international
> treaties, agreements or assurances, or participation in
> a common plan or conspiracy for the accomplishment
> of any of the foregoing;...

Waging a "weather war" would surely be a Crime Against Peace, no?

Since World War II several multilateral and regional conventions have arisen that regulate the weapons that can be used legally in armed conflicts. Restrictions on anti-personnel land mines are one example, environmental modification is another.

The International Court of Justice (ICJ) has recognized that weapons of mass destruction, which presumably would include weather weapons, are subject to international customary law. In its July 1996 ruling on nuclear weapons, "Advisory Opinion on Legality of the threat or use of Nuclear Weapons," the ICJ concluded that:

> States must never make civilians the object of
> attack and must consequently never use weapons that
> are incapable of distinguishing between civilian and
> military targets... it is prohibited to cause unnecessary
> suffering to combatants: it is accordingly prohibited
> to use weapons causing them such harm or uselessly
> aggravating their suffering. In application of that
> second principle, States do not have unlimited freedom
> of choice of means in the weapons they use.

Clearly, should any nation attempt to wage a "weather war," overtly or covertly, it would be a violation, in letter and/or spirit, of most of these international treaties and agreements. Would a nation risk it? Frankly, what's to stop them? A close reading of history shows us that in any war crimes are committed by all sides, but only the losers are ever sent to trial. What does it matter to them then? Here we see another "beauty" of covert warfare—no overt war, no messy trail after!

ENVIRONMENTAL MODIFICATIONS COVERED BY ENMOD

ENMOD is both clear and vague at the same time in delineating what signatories can and cannot do under the treaty. Article I requires that the nations which have bound themselves to this agreement (called State Parties) promise not to engage in military or any other hostile use of environmental modification techniques having widespread, long-lasting or severe effects as the means of destruction, damage or injury to any other State Party. Also, Parties agree not to assist, encourage or induce any State, group of States or international organization to engage in such activities. Any State Party that sponsored terrorist organizations would therefore be in violation of this treaty should any of the organizations they sponsor engage in "eco-type terrorism," as Secretary of Defense Cohen put it.

Some environmental manipulation, however, is specifically permitted under ENMOD. For example, armies may use herbicides or other means to denude the perimeter of military bases in order to reduce the chance of sneak attack.

While ENMOD provides protection against manipulation of the environment as a weapon of war, it does not provide protection against the environmental damage that might result from military actions, so-called "collateral damage." Consequently, intentional ruination of farmland would be considered an illegal form of warfare, whereas environmental damage resulting from a bombing campaign, for example, destruction of a chemical plant that released toxins that devastated downwind croplands, would not. This provides the "wiggle room" that the American environmental coalition objected to, and as we shall see, what the U.S. and others have used.

To continue quoting from "Addressing Environmental

121

Modification in Post-Cold War Conflict—The Convention on the Prohibition of Military or Any Other Hostile Use of Environmental Modification Techniques (ENMOD) and Related Agreements" by Susana Pimiento Chamorro and Edward Hammond:

> To violate ENMOD, an attempt to manipulate natural processes is required. The treaty does not protect states against environmental damages resulting from hostile actions. In other words, collateral damage incidental to warfare is not prohibited. Article II lists examples of environmental modification techniques and includes among them provocation of earthquakes, tsunamis, hurricanes (typhoons), disruption of ecological balance in climatic elements, change in ocean currents, and changes in the state of the ozone layer.
>
> Article I provides a definition of environmental modifications the treaty is meant to cover. That definition entails two standards, the first on the magnitude of the modification itself, i.e., the scale of damage to the environment, and the second on the intent or purpose of the modification.
>
> On magnitude, Article I establishes that for a technique to fall within the scope of ENMOD, at least one out of three criteria must be met: widespread, long lasting or severe. These criteria together are known as the "troika," and set the magnitude threshold for violation. If the threshold is met, the exception of military necessity cannot be claimed.

The definition of the *troika* was established in September 1984, when the State Parties convened in Geneva, Switzerland to hold the First Review Conference of ENMOD. The State Parties were only slightly (less than a year) behind schedule in fulfilling the mandate of Article VIII, which required them to review the convention five years after deposit of the twentieth ratification. More disappointing than their tardiness was the attendance. Only a very small number of countries had ratified the treaty after it was opened for signature in May 1977. In 1984 the number of State Parties stood at a mere 45, ten of which did not even bother to send delegates, leaving representatives from just 35 State Parties to conduct the Review Conference.

Several countries, particularly Sweden, Egypt and New Zealand pushed to expand the scope of the treaty and/or to reduce the threshold level of the *troika*. Unfortunately, in the end they were unable to do little more than have the language in the Final Declaration of the conference acknowledge the need for the troika to remain under *"continuing review and examination"* in order to assess its effectiveness, the impact of new technologies, and the views of the countries that favor an expansion of the treaty's scope.

The Conference adopted an understanding of the troika that had been previously developed by the Geneva-based Conference of the Committee on Disarmament (CCD) in the course of ENMOD's negotiation. The terms of the troika were defined as:

- Widespread: refers to the geographic area affected — the environmental modification technique used must have negatively impacted an area of several hundred square kilometers or more;
- Long lasting: the effects of the environmental modification must last for at least a period of months, or over a season;
- Severe: means the environmental modification caused serious or significant disruption or harm to human life, natural and economic resources or other assets.

Again from "Addressing Environmental Modification in Post-Cold War Conflict" by Pimiento Chamorro and Hammond:

The second standard in the definition of environmental modification relates to the intent of the modification. ENMOD only bans environmental modification for military or hostile purposes. Peaceful modifications of the environment, stemming from potential future developments of technology, are recognized as potentially desirable. "Military" refers to armed conflicts—those between State Parties (i.e. countries which have ratified ENMOD). "Hostile," a much wider concept, includes both armed and unarmed conflicts. For example, the spraying of herbicides by the United Kingdom on opium poppies in Burma without Burma's permission is a hypothetical example of a non-military hostile conflict.

Despite the best efforts of Senator Pell and the environmental lobby, development and testing of environmental modification techniques are not currently outlawed by ENMOD. The original draft of the environmental modification convention (found in US Senate Resolution 71) did contain a prohibition of "any research or experimentation directed to the development of such activity [environmental or geophysical modification] as a weapon of war." Nevertheless, development and testing were not outlawed in the final version of ENMOD. Here is another case of "wiggle room" whereby the military-industrial-academic complex can freely develop weapons without even needing the flimsy pretext of engaging in civilian science.

CRIME AND PUNISHMENT UNDER ENMOD

Again from "Addressing Environmental Modification in Post-Cold War Conflict" by Susana Pimiento Chamorro and Edward Hammond:

Who then is protected by ENMOD? There is no

question that ENMOD protects State Parties to the Convention against environmental modification techniques undertaken by other State Parties. Unfortunately, Parties have never adequately clarified the Convention's applicability to cases in which a State Party attacks a non-Party (or vice-versa). Under a restrictive interpretation of ENMOD's applicability, protection has been interpreted as an incentive to promote ratification. That is, to avoid "free riders" (those who would gain the benefits of ENMOD without having to abide by its rules), ENMOD's logic has been interpreted to mean that countries can get little protection from ENMOD without ratifying the Convention. This interpretation has been criticized by those environmentalists and leaders in the US Senate who wanted the Convention to apply to the entire international community.

Another interpretation of ENMOD grants limited protection to non-parties. This interpretation can be deduced from the wording of ENMOD's second obligation:

Article I. 2. Each State Party to this Convention undertakes not to assist, encourage or induce any State, group of States or international organization to engage in activities contrary to the provisions of paragraph 1 of this article.

Under the second interpretation, it could argued, for example, that the US is responsible for promoting massive use of chemical and biological herbicides in the Drug War and, should damage be proven, the US could be held responsible for encouraging and assisting non-parties (e.g., Colombia) and international organizations (e.g., the United Nations Drug Control Program [UNDCP]) to undertake hostile environmental modification in the form of counterinsurgency crop eradication programs. Of course, the hostility element would also have to be proven, but, in the case of Colombia at least, the US and UNDCP have argued explicitly and repeatedly that the situation is a hostile one and that anti-narcotics operations and counterinsurgency are inseparable.

While ENMOD may offer some degree of protection to non-parties, it is clear that the responsibilities imposed by ENMOD are restricted to those who have ratified the Convention. Further, ENMOD does not include a system of compensation for damages resulting from breach of its obligations. Any compensation must be sought via other international legal instruments.

Not only is there no mechanism for compensation of damages if a State Party could prove to have been violated under ENMOD, the very system to bring a complaint is heavily loaded in favor of the five States that are the permanent members of the UN Security Council, as explained by Pimiento Chamorro and Hammond:

> In ENMOD, cooperation and consultation are the preferred methods of resolving alleged violations. Any State Party which has a reason to believe that another State Party has breached the treaty may lodge a complaint with the UN Security Council. The complaint is expected to include relevant information and, if possible, evidence. Following deliberation, the Security Council — applying its standard voting procedure — is expected to decide if a violation has taken place.
>
> In addition to providing for the initiation of an investigation through the Security Council, ENMOD provides for creation of a Consultative Committee of Experts by the Secretary General at the request of any state Party. Any state Party may appoint an expert to that Committee. The committee's functions are defined in Annex 1. Primarily, the Committee is to conduct fact-finding related to the application of the

Convention and to provide its views to the Security Council. The Committee is expected to review the facts prior to any Security Council action, although the Committee cannot decide whether a violation has taken place and by whom. That power is reserved to the Security Council.

The veto power granted to the Security Council's five permanent members is, without doubt, a major constraint on the effectiveness of ENMOD since it places five states in a permanently privileged position.

LIMITATIONS OF ENMOD

Clearly, ENMOD's careful diplomatic wording and the threshold conditions set by the troika make the treaty far less effective than Senator Pell envisioned, just as the environmental lobby predicted. In this final section on it let me return to the work of Susana Pimiento Chamorro and Edward Hammond where they discuss some of the problems they identified:

The Troika: Too High a Threshold?
Critics of ENMOD have focused on the troika, arguing that it sets a threshold so difficult to cross that the treaty is practically unusable. Among these critics are Parties to the Convention such as Argentina, Mexico, Mauritius, and Turkey. The US environmental organizations that opposed US ratification of ENMOD felt that the subjective nature of the third condition — severity — could permit environmental modifications such as cloud seeding and defoliation a view that was not shared by Senate proponents of the treaty.
Unused, Lack of Political Will
During the quarter century ENMOD has existed, no state Party has been formally accused of a violation. The two review conferences held so far have stated that "the obligations assumed under article I of the

Convention had been faithfully observed by the State Parties."

Severe environmental modification techniques, however, have been used in conflicts subsequent to ENMOD's coming into force. Examples include the use of defoliants in Central America during the 1980s and the severe environmental modifications made during the Gulf War. In the latter, Iraq set fire to over 600 oil wells and targeted Kuwait's water desalinization plants by polluting the nearby sea. The UN Security Council passed a resolution condemning Iraq's invasion of Kuwait and holding it liable for damages. ENMOD, however, was not applied because Iraq is not a Party to the Convention.

Restrictions Too Limited

ENMOD does not outlaw development and testing of hostile environmental modification techniques, nor does it include verification mechanisms for identifying attempts by Parties to develop environmental modification techniques. These omissions were not accidental; at the time of ENMOD's negotiation, the USSR and the US agreed that environmental techniques could be used for peaceful purposes.

In addition to its lack of prohibition of development and testing, ENMOD does not prohibit anyone from threatening to use hostile environmental modification. Further, damage in ENMOD must be proven. Thus, difficulty arises in reconciling ENMOD — which requires after the fact scientific assessment of damages — with the Precautionary Principle, a cornerstone of environmental law whose emphasis on avoiding environmental damage is quite different than ENMOD.

Questions of damage become even more complicated when the affected ecosystems are complex and poorly understood, as is the case in many areas of the tropics containing high levels of biodiversity.

The relative lack of scientific knowledge of particular ecosystems may hamper accurate assessment of the effects of hostile environmental modification. Because ENMOD invokes its (Article I) assessment after damage is done, the true environmental extent of the damage can never be known in ecosystems not fully characterized before the damage is done.

Lack of liability system

ENMOD lacks provisions for penalizing State Parties that breach its provisions. A State Party can be held responsible, but not liable. Liability and redress, therefore, must be sought through other international legal instruments. In lieu of imposing liability, or as a mechanism to force acceptance of liability, the Security Council could decide to punish violators through the application of trade or other sanctions.

Limited number of Parties

Although many major powers are Parties to ENMOD, the overall number of ratifications is limited, particularly in the political South. To date, ENMOD has only 70 State Parties. The UN General Assembly has called for global ratification on several occasions, but without much success.

Not only does ENMOD fail to fully address these issues, so too the international community has failed to move on this.

Michel Chossudovsky is a professor of economics at the University of Ottawa. He has taught as visiting professor at academic institutions in Western Europe, Latin America and Southeast Asia; has acted as an economic adviser to the governments of several developing countries; and has worked as a consultant for numerous international organizations including the United Nations Development Programme (UNDP), the African Development Bank, the United Nations African Institute for Economic Development and Planning (AIEDEP), the United Nations Population Fund (UNFPA), the

International Labour Organization (ILO), the World Health Organization (WHO), and the United Nations Economic Commission for Latin America and the Caribbean (ECLAC).

Professor Chossudovsky is also an active member of the anti-war movement in Canada. He is currently an editor for the Centre for Research on Globalization, which is "committed to curbing the tide of 'globalization' and 'disarming' the New World Order." His article "Washington's New World Order Weapons Have the Ability to Trigger Climate Change" is a denunciation of HAARP that reads like a condensed version of my previous book on HAARP. In it he wrote:

> Despite a vast body of scientific knowledge, the issue of deliberate climatic manipulations for military use has never been explicitly part of the UN agenda on climate change. Neither the official delegations nor the environmental action groups participating in the Hague Conference on Climate Change (CO6) (November 2000) have raised the broad issue of "weather warfare" or "environmental modification techniques (ENMOD)" as relevant to an understanding of climate change.
>
> Why then did the UN—disregarding the ENMOD Convention as well as its own charter—decide to exclude from its agenda climatic changes resulting from military programmes?

As mentioned above, environmental modification techniques have been used in conflicts since the ratification of ENMOD—in the War on Drugs, both Gulf Wars and in Kosovo, to name a few. ENMOD violations have occurred, and have been broadcast in (dying?) color on the world's news channels for a couple of decades. Concurrently rumors of covert geophysical warfare grow at an alarming rate. Strange weather and freak earthquakes have people around the world frightened and asking questions. Yet no aggrieved State Party (signatory to ENMOD or otherwise) has stepped forward with a complaint. And all the while the military-industrial-academic complexes of the world's great powers continue to develop plans and techniques for the next weather war. Where is it all heading?

AIR FORCE 2025: OWNING THE WEATHER

On 17 February 2002 ABC News ran a segment titled *Weather As A Weapon?* The sole US Air Force official they interviewed for the piece was Director of Weather Brigadier General Fred Lewis. He told ABC "We want to anticipate and exploit the weather, not modify it."

I hope you can excuse me for not believing that General Lewis is being entirely forthcoming in that statement. I am quite sure that The United States Air Force fully realizes the need to adhere to the letter of ENMOD, while at the same time I am quite sure that they would like to skirt its restraints as much as possible. What General Lewis carefully did not say was that military environmental modification is not controlling the weather, but *modifying the atmosphere!*

One clear example of how to both keep and break the ENMOD treaty at the same time can be seen in the previously mentioned "Weather as a Force Multiplier: Owning the Weather in 2025."

Force multiplier is a military term referring to a factor that dramatically increases (hence "multiplies") the combat effectiveness of a military force. Some common force multipliers are: technology, morale, geographical features, training, strength of numbers, and, as we have been discussing, the weather. General Lewis' comment that the Air Force wants to "anticipate and exploit" the weather clearly means that they fully intend to use it as a force multiplier within the restrictions of ENMOD. But *2025* goes well beyond that.

"Weather as a Force Multiplier: Owning the Weather in 2025" was prepared by Col. Tamzy J. House, Lt. Col. James B. Near, Jr., LTC William B. Shields (USA), Maj. Ronald J. Celentano, Maj. David M. Husband, Maj. Ann E. Mercer, and Maj. James E. Pugh. They presented it on 17 June 1996 to the *Air Force 2025 committee*. It was "a study designed to comply with a directive from the chief of staff of the Air Force to examine the concepts, capabilities, and technologies the United States will require to remain the dominant air and space force in the future."

The opening Disclaimer states that: "The views expressed in this report are those of the authors and do not reflect the official policy or position of the United States Air Force, Department of Defense, or the United States government." However, they also state that the weather modification capabilities described in the report are consistent with the "operating environments and missions" of the Air Force's long range planning office. "Weather as a Force Multiplier: Owning the Weather in 2025" is probably a good indicator of where the military plans, wants or at least feels it needs to go with weather modification technology in the years ahead.

The Executive Summary of 2025 begins:

> In 2025, US aerospace forces can "own the weather" by capitalizing on emerging technologies and focusing development of those technologies to war-fighting applications. Such a capability offers the war fighter tools to shape the battlespace in ways never before possible. It provides opportunities to impact operations across the full spectrum of conflict and is pertinent to all possible futures.
>
> A high-risk, high-reward endeavor, weather-modification offers a dilemma not unlike the splitting of the atom. While some segments of society will always be reluctant to examine controversial issues such as weather-modification, the tremendous military capabilities that could result from this field are ignored at our own peril. From enhancing friendly operations or disrupting those of the enemy via small-scale tailoring of natural weather patterns to complete dominance of global communications and

counterspace control, weather-modification offers the war fighter a wide-range of possible options to defeat or coerce an adversary. Some of the potential capabilities a weather-modification system could provide to a war-fighting commander in chief (CINC) are listed in table 1.

Technology advancements in five major areas are necessary for an integrated weather-modification capability: (1) advanced nonlinear modeling techniques, (2) computational capability, (3) information gathering and transmission, (4) a global sensor array, and (5) weather intervention techniques. Some intervention tools exist today and others may be developed and refined in the future.

Current technologies that will mature over the next 30 years will offer anyone who has the necessary resources the ability to modify weather patterns and their corresponding effects, at least on the local scale. Current demographic, economic, and environmental trends will create global stresses that provide the impetus necessary for many countries or groups to turn this weather-modification ability into a capability.

Appropriate application of weather-modification can provide battlespace dominance to a degree never before imagined. In the future, such operations will enhance air and space superiority and provide new options for battlespace shaping and battlespace awareness. "The technology is there, waiting for us to pull it all together;" in 2025 we can "Own the Weather."

"Chapter 4, Concept of Operations" gets to the heart of the matter. Here are a few particularly juicy paragraphs pulled from the text:

The essential ingredient of the weather-modification system is the set of intervention

133

techniques used to modify the weather. The number of specific intervention methodologies is limited only by the imagination, but with few exceptions they involve infusing either energy or chemicals into the meteorological process in the right way, at the right place and time. The intervention could be designed to modify the weather in a number of ways, such as influencing clouds and precipitation, storm intensity, climate, space, or fog.

International agreements have prevented the US from investigating weather-modification operations that could have widespread, long-lasting, or severe effects. However, possibilities do exist (within the boundaries of established treaties) for using localized precipitation modification over the short term, with limited and potentially positive results. These possibilities date back to our own previous experimentation with precipitation modification. As stated in an article appearing in the *Journal of Applied Meteorology:*

> *Nearly all the weather-modification efforts over the last quarter century have been aimed at producing changes on the cloud scale through exploitation of the saturated vapor pressure difference between ice and water. This is not to be criticized but it is time we also consider the feasibility of weather-modification on other time-space scales and with other physical hypotheses.*

The desirability to modify storms to support military objectives is the most aggressive and controversial type of weather-modification. The damage caused by storms is indeed horrendous. For instance, a tropical storm has an energy equal to 10,000 one-megaton hydrogen bombs, and in 1992 Hurricane Andrew totally destroyed Homestead AFB, Florida, caused the evacuation of most military aircraft in the southeastern US, and resulted in $15.5 billion of damage.

Modification of the near-space environment is crucial to battlespace dominance. General Charles Horner, former commander in chief, United States space command, described his worst nightmare as "seeing an entire Marine battalion wiped out on some foreign landing zone because he was unable to deny the enemy intelligence and imagery generated from space." Active modification could provide a "technological fix" to jam the enemy's active and passive surveillance and reconnaissance systems. In short, an operational capability to modify the near-space environment would ensure space superiority in 2025; this capability would allow us to shape and control the battlespace via enhanced communication, sensing, navigation, and precision engagement systems.

Manipulation and modification of the ionosphere is taken up in depth in this section of "Weather as a Force Multiplier: Owning the Weather in 2025." The passages quoted below seem to have been written with HAARP in mind, as we shall see in greater depth in the next chapter.

Modification of the ionosphere to enhance or disrupt communications has recently become the subject of active research. According to Lewis M. Duncan, and Robert L. Showen, the Former Soviet Union (FSU) conducted theoretical and experimental research in this area at a level considerably greater than comparable programs in the West.

...Modification of the ionosphere is an area rich with potential applications and there are also likely spin-off applications that have yet to be envisioned.

...The major disadvantage in depending on the

ionosphere to reflect radio waves is its variability, which is due to normal space weather and events such as solar flares and geomagnetic storms. The ionosphere has been described as a crinkled sheet of wax paper whose relative position rises and sinks depending on weather conditions. The surface topography of the crinkled paper also constantly changes, leading to variability in its reflective, refractive, and transmissive properties.

...An artificial ionospheric mirror (AIM) would serve as a precise mirror for electromagnetic radiation of a selected frequency or a range of frequencies. It would thereby be useful for both pinpoint control of friendly communications and interception of enemy transmissions.

The ionosphere could potentially be artificially charged or injected with radiation at a certain point so that it becomes inhospitable to satellites or other space structures. The result could range from temporarily disabling the target to its complete destruction via an induced explosion. Of course, effectively employing such a capability depends on the ability to apply it selectively to chosen regions in space.

In contrast to the injurious capability described above, regions of the ionosphere could potentially be modified or used as-is to revitalize space assets, for instance by charging their power systems. The natural charge of the ionosphere may serve to provide most or all of the energy input to the satellite. There have been a number of papers in the last decade on electrical charging of space vehicles; however, according to one author, "in spite of the significant effort made in the field both theoretically and experimentally, the vehicle charging problem is far from being completely understood." While the technical challenge is considerable, the potential to harness electrostatic energy to fuel the satellite's power cells would have a high payoff, enabling service life extension of space assets at a relatively low cost. Additionally, exploiting the capability of powerful HF radio waves to accelerate electrons to relatively high energies may also facilitate the degradation of enemy space assets through directed bombardment with the HF-induced electron beams. As with artificial HF communication disruptions... the degradation of enemy spacecraft with such techniques would be effectively indistinguishable from natural environment effects. The investigation and optimization of HF acceleration mechanisms for both friendly and hostile purposes is an important area for future research efforts.

"Chapter 5: Investigation Recommendations. How Do We Get There From Here?" concludes with:

Even today's most technologically advanced militaries would usually prefer to fight in clear weather and blue skies. But as war-fighting technologies proliferate, the side with the technological advantage will prefer to fight in weather that gives them an edge. The US Army has already alluded to this approach in their concept of "owning the weather." Accordingly, storm modification will become more valuable over time. The importance of precipitation modification is also likely to increase as usable water sources become more scarce in volatile parts of the world.

As more countries pursue, develop, and exploit increasing types and degrees of weather-modification technologies, we must be able to detect their efforts and counter their activities when necessary. As depicted, the technologies and capabilities associated with such a counter weather role will become increasingly important.

The world's finite resources and continued needs will drive the desire to protect people and property and more efficiently use our crop lands, forests, and range lands. The ability to modify the weather may be desirable both for economic and defense reasons.

The lessons of history indicate a real weather-modification capability will eventually exist despite the risk. The drive exists. People have always wanted to control the weather and their desire will compel them to collectively and continuously pursue their goal. The motivation exists. The potential benefits and power are extremely lucrative and alluring for those who have the resources to develop it. This combination of drive, motivation, and resources will eventually produce the technology. History also teaches that we cannot afford to be without a weather-modification capability once the technology is developed and used by others. Even if we have no intention of using it, others will. To call upon the atomic weapon analogy again, we need to be able to deter or counter their capability with our own. Therefore, the weather and intelligence communities must keep abreast of the actions of others.

Weather-modification is a force multiplier with tremendous power that could be exploited across the full spectrum of war-fighting environments. From enhancing friendly operations or disrupting those of the enemy via small-scale tailoring of natural weather patterns to complete dominance of global communications and counter-space control, weather-modification offers the war fighter a wide-range of possible options to defeat or coerce an adversary. ... But, while offensive weather-modification efforts would certainly be undertaken by US forces with great caution and trepidation, it is clear that we cannot afford to allow an adversary to obtain an exclusive weather-modification capability.

In Keith Harmon Snow's "Out of the Blue: Black Programs, Space Drones & The Unveiling of U.S. Military Offensives in Weather as a Weapon" he comments on *AF2025* and US Air Force Director of Weather Brigadier General Fred Lewis' quote that "We want to anticipate and exploit the weather, not modify it" saying:

> The entire subject of weather warfare revolves around "plausible deniability" and the capacity of elite decision makers to "plausibly deny" that such technologies exist (just as assassinations were not committed, coups were not fomented, massacres were not perpetrated, "disappeared" people were not disappeared). Because proof of secret operations is highly classified, hence invisible, the unverifiable accusations are answered with plausible denials and we are left to depend on the basic goodness and integrity of leaders — whom are otherwise insane, and routinely lie through their teeth, while looking you straight in the eye.

The statement by Brig. General Fred Lewis is contradicted, in its most simple form, by the obvious fact that all branches of the U.S. military and security apparatus rely on sophisticated SIGINT (SIGnals INTelligence), COMINT (COMmunications INTelligence), EW (Electronic Warfare), C4ISR/T (Command, Control, Communications, Computers, Intelligence, Surveillance and Reconnaissance / Tracking) technologies whose entire mission and purpose can be, and often has been, compromised, neutralized or entirely defeated by hostile (but natural) weather conditions in the battlespace environment.

Question: Would the defense establishment ignore technologies with the potential to decisively defeat the "no-win" military compromises and losses due to an unpredictable and cranky old mother nature?

Answer: They would not.

The statement by Brig. General Fred Lewis is further contradicted by the obvious military thrusts to develop capabilities that maximize stealth and, simultaneously, minimize risk to U.S. troops, and the propensity, again, well documented, to use clandestine operations premised, again, on "plausible denial." In light of these major policy and field objectives, the existence of an entire spectrum or portfolio of ENMOD technologies is both plausible and certain. Said differently, it is irrational and unlikely and naïve and unreasonable to suppose the absence of these technologies given the grossly offensive actions that have been consistently demonstrated by the U.S. military, scientific and intelligence establishment over the past five decades.

Owning the Weather in 2025 …confirms the offensive interests the U.S. Air Force has in "owning and controlling" weather—as a weapon. Numerous citations and references from the report itself reveal that military analysts and scientists have been working on weather modification issues in some capacities. (Projected ENMOD capabilities from the report are delineated in Table One.)

Keith goes on to quote several key paragraphs from *AF2025*. One such was:

A number of methods have been explored or proposed to modify the ionosphere, including injection of chemical vapors and heating or charging via electromagnetic radiation or particle beams (such as ions, neutral particles, x-rays, MeV particles, and energetic electrons). It is important to note that many techniques to modify the upper atmosphere have been successfully demonstrated experimentally. Ground-based modification techniques employed by the FSU include vertical HF heating, oblique HF heating, microwave heating, and magnetospheric modification. Significant military applications of such operations include low frequency (LF) communication production, HF ducted communications, and creation of an artificial ionosphere. Moreover, developing countries also recognize the benefit of ionospheric modification: "in the early 1980's, Brazil conducted an experiment to modify the ionosphere by chemical injection."

Keith comments on this saying:

> First note that this latter paragraph admits that both the U.S.S.R. and Brazil—the latter as recent as the 1980s—have performed ENMOD experiments on the earth's ionosphere. This curious admission raises at least two questions to ponder prior to exploring the true U.S. involvement in the ENMOD arena: [1] Would the U.S. DOD stand idly by and twiddle their missiles while other nations develop potentially revolutionary lethal technologies that might be then used against the U.S.? And [2]: Were not the U.S.S.R. and Brazil's ENMOD experiments in contravention of the International ENMOD treaty of 1977?

In my previous book on this subject, *HAARP: The Ultimate Weapon of the Conspiracy*, I examined how HAARP could be a part of such a futuristic scenario as described in *AF2025* in general and how it has been used for ionospheric modification in particular. We will explore this in depth in the next chapter.

The authors of *2025* insist that we cannot allow potentially hostile governments (and vague asymmetric threats, i.e. terrorists) to develop this technology and not do like-wise. Perhaps they are right. Clearly this should be a significant national debate. Why isn't it? Could it be because doing so would reveal what our government is actually doing—developing full-on EnMod warfare capabilities?

Unfortunately, very little from the black world of secret ops has leaked out to us to date. But, from what has appeared in the publicly available media we can surmise a great deal about

141

what has been developed and where this technology, and its use, is headed.

 Let us take up HAARP next as one clear indicator of how the military-industrial-academic complex has kept this a secret in plain sight.

Nikola Tesla (1856-1943)

Top: Tesla in the background with Mark Twain, circa 1895. Bottom: Tesla on the cover of *The Electrical Experimenter*

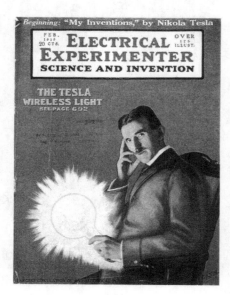

Tesla's broadcasting tower as featured on his stationary, circa 1917.

Weather warfare Tesla-style.
Top: Tesla's stationary for his
company, circa 1920. Below:
His tower partially built in 1918.

Nikola Tesla: The father of weather warfare?

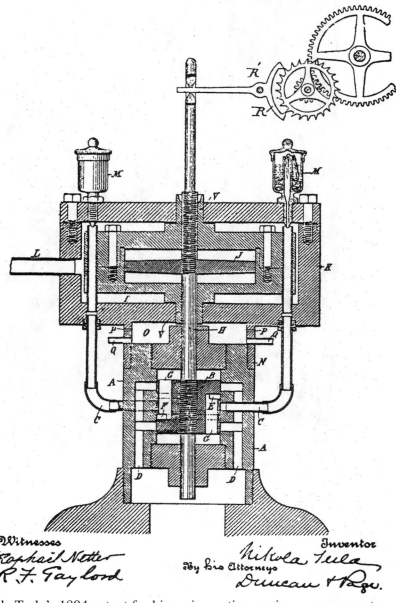

(No Model.)

N. TESLA.
RECIPROCATING ENGINE.

No. 514,169.

Patented Feb. 6, 1894.

Witnesses
Raphael Netter
R. F. Taylord

Inventor
Nikola Tesla
By his Attorneys
Duncan & Page.

Nikola Tesla's 1894 patent for his reciprocating engine, a precursor to his "Earthquake Machine."

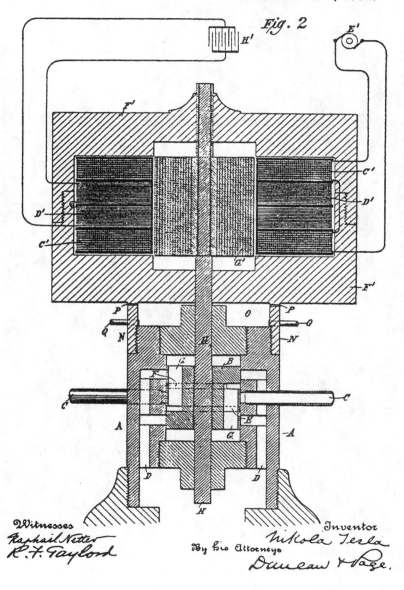

Nikola Teslä's 1894 patent for his electric generator, a component to his "Earthquake Machine."

Tesla's wireless transmission tower in action, sending power to electrical airships in this illustration from *Radio News,* December, 1925.

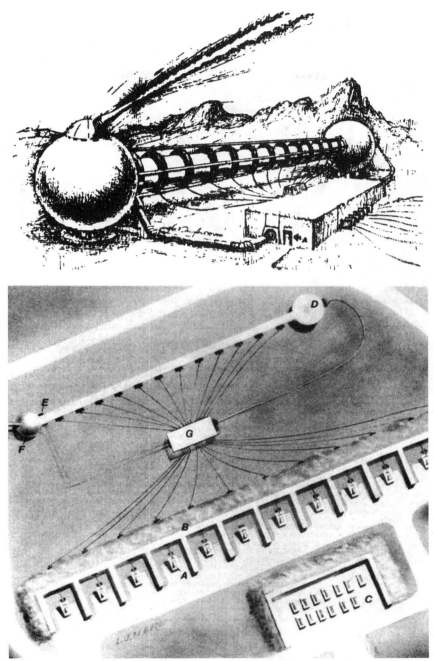

Top: Hal Crawford's drawing of the suspected Russian beam weapon installation near Semipalatinsk. Below: Overview of the installation.

151

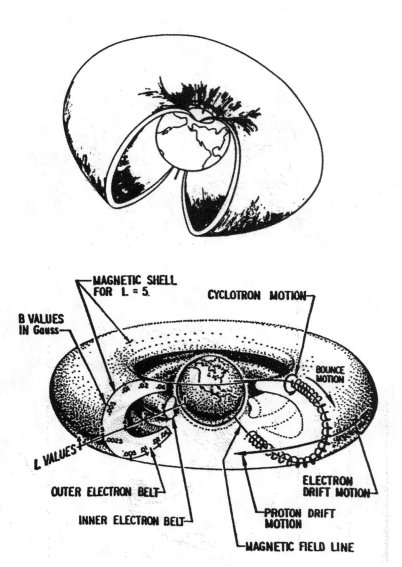

Top: The Earth's magnetic shell. Below: Elements of the Earth's magnetic shell that could be utilized by the HAARP project.

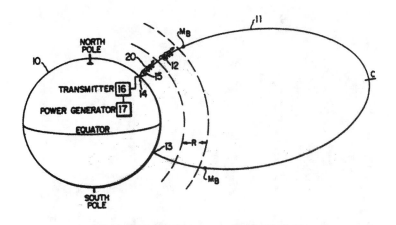

Top: Bernard Eastlund's 1987 patent (number 4,686,605) depicting a method for altering a selected region of the earth's surface. Below: A close-up of the HAARP antenna array in Alaska.

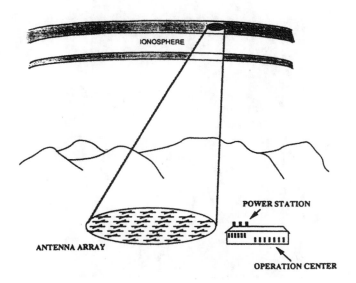

Top: A close-up of the HAARP antenna array in Alaska. Bottom: Diagram of how the array can heat the atmosphere.

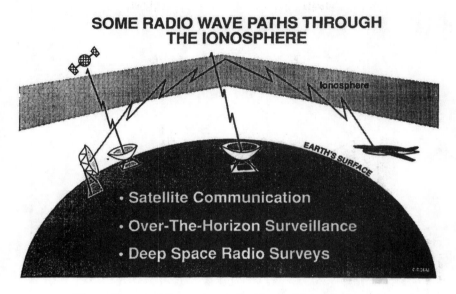

Top: Some radio wave paths through the Ionosphere and how they can be used.

CHAPTER FIVE

HAARP

The HAARP Interactive Ionospheric Research Observatory is a major Arctic facility for the study of upper atmospheric and solar-terrestrial physics and for Radio Science and Communications research. Among the instruments included at the facility are a high power, high-frequency (HF) phased array radio transmitter, numerous radio frequency and optical research instruments capable of observing and monitoring the complex auroral ionosphere, and site infrastructure to support research activities.

The above is from the University of Alaska at Anchorage's official HAARP website. The Ionospheric Research Instrument (IRI) is the high-frequency phased array radio transmitter mentioned in the paragraph above. It is the heart of the High-frequency Active Auroral Research Program (HAARP).

The IRI is a field of towers, each 72 feet tall with two crossed dipole antennas at the top of each tower. These antennas are for broadcasting in the High Frequency (HF) range, which is an audible range below the lowest ranges currently used for radio communications (below shortwave). Slung beneath the towers is a wire grid for reflecting upward any radio emissions from the antennas that heads toward the ground. Beneath the mesh are small buildings called shelters for the radio transmitters. All of these elements are linked together to act as one giant transmitting antenna. HAARP began at the end of the 1980s and may have been completed in 2005, its completion being announced in 2006.

As you can imagine, a lot has changed with the HAARP

project and with those watching it since news of it first reached the public in the early 1990s. For me, one of the more remarkable changes has been that Dr. Nick Begich and I have swapped positions since our respective books on HAARP appeared. Dr. Begich and Jeane Manning self-published the first book on HAARP, *Angels Don't Play This HAARP: Advances In Tesla Technology*, in 1995. At that time they were convinced that HAARP was a ground-based Star Wars weapons system, an outgrowth of the Strategic Defense Initiative.

When I learned of the project, through their book and Internet scuttlebutt, I was shocked and frightened to learn that HAARP was designed to bathe the world in extremely low radio frequencies (ELF) at exactly the same frequency that the human brain works at—how it does this will be explained shortly. After twenty years of studying mind control technologies my first thought was that if HAARP were exactly what the program's personnel said it was, it could inadvertently cause mental dysfunction across whole continents without the scientists involved even being aware they were doing it. Worse, if they were using it intentionally for covert purposes, HAARP could be the ultimate global mind control device. That is partly why my book, the second on HAARP published in 1998, was called *HAARP: The Ultimate Weapon of the Conspiracy*.

As I mentioned in my Introduction to this book Begich and Manning also saw that possibility. As they wrote in "Vandalism In The Sky?" quoting from Brzezinski on a proposal from Dr. Macdonald:

> Political strategists are tempted to exploit research on the brain and human behavior. Geophysicist Gordon J. F. MacDonald—specialist in problems of warfare—says [an] accurately-timed, artificially-excited electronic stroke "…could lead to a pattern of oscillations that produce relatively high power levels over certain regions of the Earth... In this way, one

158

could develop a system that would seriously impair the brain performance of very large populations in selected regions over an extended period..."

A lot has happened since the publication of our books. Indeed, now, nearly a decade later, Nick and I have traded places: he is convinced that HAARP *is* a worldwide mind control device as suggested by Dr. MacDonald and I think it is a ground-based Star Wars weapons system — and a whole lot more!

WHAT IS HAARP?

The High-frequency Active Auroral Research Program (HAARP) is the largest and most powerful United States Department of Defense (DoD) facility of its kind in the world — just what kind of facility it is, however, has been the subject of heated debate for many years.

Located in southeastern Alaska, HAARP is a field of antennas so constituted that it works as one giant antenna. Moreover, it is the world's largest radio broadcasting station, now with an effective radiated power of 3.6 million watts — more than 72,000 times more powerful than the largest legal commercial radio station (50,000 watts) in the United States!

But, HAARP's broadcasts are not intended for human ears. Its purpose is to inject all that radio frequency energy into a "spot" at the very top of the atmosphere, in a region called the ionosphere. That spot, according to some documents published by the project, is about 12 miles across, by about 2 miles deep, by about 50 or 90 miles up, depending on which mode they are using.

Why do they want to inject all that energy into the ionosphere? Doing so mimics the way the sun injects energy into the same region of the atmosphere. By artificially recreating natural phenomena they hope to better understand the processes and problems caused by solar flares and such and to control and create various ionospheric conditions. All of the official statements from military and academic sources have insisted that HAARP is a pure science research station, one intended to increase our understanding of the upper atmosphere, and nothing more. Detractors are not convinced.

Some researchers think HAARP may be an over-the-horizon type radar ... or it may exist to destroy incoming ICBMs ... or perhaps to fry out the electronics of enemy spy satellites ... or it might be used to disrupt enemy radio communications. Researchers think it might be intended to do these things because those abilities are all described in the original 12 patents that the HAARP antenna array is based on (the first of which was reproduced at the back of my 1998 book, all of which are listed in Appendix B of this book). The folks running HAARP deny any connection between HAARP and those patents, however. This, naturally, has opened them up to charges of cover-up and conspiracy.

Once one gets the idea that the government is lying about something, the sky is pretty much the limit for speculations about what it is they are lying about, and why. And with HAARP the sky is just the beginning!

HAARP has been in use, at varying output strengths, since 1995. Even before HAARP began running its first tests researchers were looking into what this technology might be capable of. Much of what they came up with sounds like pure science fiction. Some reported that HAARP would be used to control the weather. Others predicted that it would be used to set off volcanoes, or trigger earthquakes using electromagnetic waves. Still others saw its potential for use in "winning the hearts and minds" of target populations by beaming emotions or commands directly into people's heads. Some speculated that it would be used by the New World Order to take over the world by projecting holographic images into the sky while beaming thoughts directly into our minds, telling us to accept the new "god" of their design (under the purported *Project Blue Beam*, which is hopefully

160

no more than an Urban Legend). Still others thought it was part of a planetary defense system to protect us from invading aliens from outer space!

I began writing *HAARP: The Ultimate Weapon of the Conspiracy* with the intent to sort out truth from fiction in the many claims about what HAARP could or could not do. I quickly learned that HAARP was but the tip of a very nasty iceberg. I soon learned more about the downside of electromagnetic (radio frequency) technologies than I ever wanted to know. And that things are far worse in the military-industrial-academic complex than even I, as a "peacenik" from the '60s, imagined.

That book became an examination of two major questions: what is the fundamental flaw in how we fund and conduct science, and how can a democracy defend itself from its own military and covert intelligence agencies? I am not sure I ever managed to state those questions in so many words in the text, though. In writing, sometimes you don't know exactly what you are doing until after you've done it, stand back a bit, and go "oh!" (and occasionally "D'oh!")

Some people have the mistaken idea that since it is a research project funded and overseen by the military, HAARP must be all secret and hush-hush. That is only partly the case. The military wants us to believe that HAARP is nothing more than a simple civilian science project, so they have done the best they can to make it look civilian. Officially there is nothing secret or classified about it. But the military has many ways to keep secrets.

One way to keep documents from the public is to simply say they were never published, which happened with some early key HAARP documents, as Dr. Begich related in *Angels*. Another way is for them to be owned by private corporations who do not have to share their secrets with the public. HAARP began as "intellectual property" (patents) owned by a private corporation, APTI, then a wholly owned subsidiary of ARCO, the oil giant. APTI was granted the primary contract to build HAARP simply because it was the only one who could, because it held the patents (which HAARP officials deny their project is based on!). Soon after construction began APTI was sold to E-Systems, which in turn was bought by Raytheon, one of America's largest defense contractors, in April, 1995. Raytheon in turn sold their E-Systems Division to BAE Systems, one of the world's largest arms dealers. With each change of ownership of the company that owned the patents also went the HAARP construction contract, empirical proof that HAARP is based on those patents.

Another way the military can keep secrets is through "compartmentalization." This is an ancient military ploy whereby each individual or unit involved in an operation only knows what they need to, to perform their role. I am convinced that HAARP is a compartmentalized operation. Students and professors from the University of Alaska do most of the "grunt" work at the HAARP site. I have no doubt that they are honest people who sincerely believe that they are only engaged in ionospheric research. On the other hand, I believe that they are only told what they need to know, and are kept blissfully ignorant of the covert operations ("black ops") side of the project (if there is one, of course). This would give them, in the modern lexicon of spin, "plausible deniability."

HAARP is designed for its experiments to be conducted remotely — that is, the people running the experiments can be anywhere in the world using the Internet and the like to control what happens in Alaska. During these remotely directed "campaigns" the guys at HAARP are there to turn on the equipment and watch the dials and gauges to ensure that everything is working properly, but the actual instructions are coming from elsewhere. I see in this a clear potential for "black ops" to take place without the civilians at HAARP having a clue that it is happening — but then, I have a suspicious mind…

HAARP IN THE MEDIA

As mentioned above, the first book about HAARP was *Angels Don't Play This HAARP: Advances in Tesla Technology* by Nick Begich, Ph. D., and Jeane Manning. Dr. Begich is the eldest son of the late United States Congressman from Alaska, Nick Begich, Sr., and political activist Pegge Begich. He is a past-president of the Alaska Federation of Teachers. He received his doctorate in traditional medicine from The Open International University for Complementary Medicines of Sri Lanka in November of 1994. Jeane Manning is an Alaska-born Canadian freelance journalist who since 1981 has traveled throughout North America and Europe to report on new-energy technologies.

When *Angels* burst on the scene in the fall of 1995 it ignited a firestorm of controversy on the Internet and in talk radio circles. A few mainstream publications, like the magazine *Popular Science*, also picked up the story.

At that time I was working closely with the late Jim Keith, author of *Mind Control, World Control* and a dozen other books. He was then riding a wave of success with his first few books, including two for IllumiNet Press: *The Gemstone File* and *Black Helicopters over America: Strikeforce for the New World Order*, which had become a minor best seller. In the spring of 1996 Jim got a call from the owner and publisher of IllumiNet, Ron Bonds (who, like Jim, later died under suspicious circumstances). Ron told Jim that HAARP was the hottest thing on the Internet and asked if he could write a book on the subject. Jim's response was that he was contracted to write four more books for other publishers and so was too swamped to do it, but he had a friend — me.

I spent the summer of 1996 writing the first draft. My title for the book was *HAARP: The Ultimate Weapon?* But *conspiracy theory* was "flavor of the month" and Ron insisted that the book had to have "conspiracy" in the title. That's how "of the Conspiracy" got tacked on, making it *HAARP: The Ultimate Weapon of the Conspiracy*. I am still a bit embarrassed by it, to be honest.

Sometimes I half-jokingly call myself a "recovering" liberal. I was a civil rights activist in the early 1960s, and an anti-war activist in the late '60s. As the war in Vietnam wound down I became an environmental activist. Around 1976 I read Gary Allen's *None Dare Call It Conspiracy* and Walter Bowart's *Operation: Mind Control*. Those set me off in a whole new direction. I began researching, and later writing about the New World Order and the "open conspiracy" to create a single, one-world government.

I became intimately familiar with the so-called "anti-government" American Patriot Movement (the unorganized militia movement isn't against the legitimate government, only against *corrupt* government, but that's a different issue...) through working with Jim Keith. I assisted in researching and editing several of his books and I also helped out with his magazine, *Dharma Combat: The Magazine of Spirituality, Reality and Other Conspiracies*, where I worked as Managing Editor and Art Director (and contributed articles, poetry and a comic strip under the penname *jarod o'danu*). Writing about HAARP, then. became the perfect vehicle to combine my environmental, anti-war and anti-globalist sentiments.

I submitted the rough draft for that book to Ron Bonds in September of 1996. Ron, however, hated it and cancelled the project! I sat on the manuscript for about a year, then remembered David Hatcher Childress, owner/president of Adventures Unlimited Press. It turned out David was looking for a book on HAARP and leapt at the chance to publish it. That was in December of 1997. Four rewrites (to improve readability) and three months later I sent David the final manuscript in February of 1998.

After *HAARP* hit the stands in August of that year other books, wholly or partly dealing with HAARP, began to appear. By some strange coincidence the majority appeared in the year 2000. These included: *Secrets of Cold War Technology: Project HAARP and Beyond*, by Gerry Vassilatos; *The Lost Journals of Nikola Tesla: HAARP – Chemtrails and The*

Secret of Alternative 4, by Tim Swartz and Timothy Beckley; *Planet Earth: The Latest Weapon of War* by Dr. Rosalie Bertell; and Nick Begich's follow-up book, co-authored with the late James Roderick, *Earth Rising: The Revolution, Toward a Thousand Years of Peace*. HAARP is also mentioned in *The Giza Death Star* by Joseph P. Farrell, which appeared in 2001. In 2002 HAARP likewise turns up in *Atlantis, Alien Visitation & Genetic Manipulation* by Michael Tsarion.

Scores of articles have appeared on the Internet and in a variety of publications, as well. Not all of those articles were written by tinfoil hat wearing paranoids, either! Numerous scientists, many with Ph.D.s, have written on HAARP. Among them is the above mentioned Dr. Rosalie Bertell. She has written extensively about the use of nuclear weapons, in particular depleted uranium, and has authored two books, *No Immediate Danger: Prognosis for a Radioactive Earth* and *Planet Earth, The Latest Weapon of War*. The latter has a lengthy section on HAARP.

Dr. Bertell has a doctorate in biometrics and has worked in the field of environmental health since 1969. She has been involved in the founding of several organizations, including the International Institute of Concern for Public Health in Toronto, Canada, of which she is President. She led the Bhopal and Chernobyl Medical Commissions, has undertaken collaborative research with numerous organizations and is the recipient of the Right Livelihood Award, the World Federalist Peace Prize, the United Nations Environment Programme (UNEP) Global 500 Award and five honorary doctorates. Dr. Bertell is also a member of a Roman Catholic religious congregation, the Grey Nuns of the Sacred Heart.

Her article *Background on the HAARP Project* was published by Dr. Nick Begich's Earthpulse Press in November 1996, and has since appeared in many places on the web. In *Planet Earth, The Latest Weapon of War* she commented on Secretary Cohen's statement that "others are engaging even in an eco-type of terrorism whereby they can alter the climate, set off earthquakes, volcanoes remotely through the use of electromagnetic waves," saying "the Military has a habit of accusing others of having capabilities they already hold!"

HAARP also began turning up in fiction. It appeared in an early Tom Clancy young adult novel where terrorists take over HAARP and threaten to "mind control" the world. It was mentioned in several American TV shows, including an episode of the American Broadcasting Company's TV series *Seven Days* where it was erroneously said to be located in Turkey.

In the 2003 major Hollywood movie *The Core* there is a secret government project to create a weapon that could create earthquakes on demand. A test of the weapon accidentally stops the Earth's core from rotating, threatening to end life on this planet. On the DVD of that film one can hear in the commentaries where Jon Amiel, the director, matter-of-factly states that the idea for the film was based on HAARP.

More recently (2004) in *H. A. A. R. P. 's Fury*, a Bryson McGann novel by William Beck, fictional terrorists use HAARP to create incredibly destructive weather.

HAARP SCIENCE

In the *Executive Summary* of a report entitled "Applications and Research Opportunities Using HAARP" issued by a scientific committee sponsored by the Air Force's Phillips Laboratory and the Office of Naval Research (and released by the Technical Information Division, Naval Research Laboratory, Washington, DC, in 1995), we read:

> The Sun controls and shapes the three major regions of Geospace—the magnetosphere, the ionosphere, and the atmosphere. These regions, rather than being isolated, interact with each other and form a chain that connects the Earth to the Sun through the atmosphere and the solar terrestrial environment. Disturbances originating at the Sun spread through this chain via the solar wind and solar radiation. They ultimately

influence our weather, our climate, and even our communications. The clouds and the Earth's surface play a critical role in this chain, which is finely tuned to be in a delicate equilibrium with life on Earth. The atmosphere and the ionosphere are the geospace regions closest to the surface of the Earth. The lowest regions of the neutral atmosphere — the troposphere, the stratosphere, and the mesosphere — are critical in controlling the global temperature of the Earth and in filtering the harmful effects of the solar radiation. The next layer — the ionosphere — starts at about 60 – 70 km altitude and contains a significant fraction of electrically charged particles. Because charged particles are subjected to electric and magnetic forces the ionosphere has a uniquely important role within the overall solar-terrestrial system. It couples to the magnetosphere and heliosphere above by electric forces and to the stratosphere below by conventional atmospheric dynamic forces. It is a region that supports and controls electric currents and potentials ranging up to a million amperes and hundreds of kilovolts, respectively.

The presence of charged particles in the ionosphere controls the performance of many military and civilian systems using electromagnetic waves. On the low frequency end (VLF/ULF/HF) refection of radio waves by the ionosphere allows for worldwide communications and Over-the-Horizon (OTH) radar operation. On the higher frequency end (VHF/UHF) transionospheric propagation is a ubiquitous element of numerous civilian and military communication, surveillance, and remote sensing systems. Paths linking satellites to the ground cross the ionosphere, and the system performance is often critically dependent on the state and structure of the ionosphere in the vicinity of these paths.

In the February 1990 report from the Air Force Geophysics Laboratory and the Office of Naval Research, "HF Active Auroral Research Program Joint Services Program Plans And Activities," we read the previously introduced quote:

> The heart of the program will be the development of a unique ionospheric heating capability to conduct the pioneering experiments required to adequately assess the potential for exploiting ionospheric enhancement technology for DOD purposes.

Let's break that down...

"...development of a unique ionospheric heating capability..." Technically HAARP is a type of device called an ionospheric heater because injecting all that radio frequency energy into the ionosphere heats it up — by several thousand degrees! The big difference between HAARP and the dozen or so other ionospheric heaters in the world is that HAARP, based on the APTI patents, is a uniquely designed phased antenna array. This phasing, or sequencing of the firing of the transmitters/antenna field allows for the focusing ability that sets HAARP apart from its peers. If used for over-the-horizon surveillance it would also have made HAARP a violation of the ABM Treaty, which was still in effect when work on HAARP began, and may be the real reason for the military's calling HAARP a civilian science project in the first place.

"...ionospheric enhancement technology..." Heating the atmosphere changes it, so you can make it do things. Because of convection and other factors you can literally shape the ionosphere, controlling how and where it bounces radio waves. Also, the amount of heat HAARP is capable of generating literally blows the molecules of the air apart. You have to love the use of the word *enhancement* — only the military would think that breaking something makes it better! This heating to the point where the molecules are blown apart causes it to give off a "scream" of extremely low.

168

frequency (ELF) radio waves that penetrate deep into the earth and deep into the seas. The project was initially funded specifically to do this: to use this ionospherically generated ELF to communicate with deeply submerged submarines and to develop something called earth penetrating tomography (EPT) to target and monitor enemy underground bases for the manufacture and launch of weapons of mass destruction (WMDs).

Earth penetrating tomography is a brand new science, one that did not exist before 1980. It has been described as being something like a "dirt radar" allowing scientists to see underground. After successful EPT tests with another ionospheric heating facility in Alaska proved that ionospherically generated ELF could be used to locate underground targets, the U.S. Senate insisted that HAARP's primary mission would be to develop EPT capability to detect and monitor underground WMD facilities as part of the United States' counterproliferation efforts. Curiously, despite this Senatorial mandate no official announcement from any organization associated with HAARP operations has ever said that it has been used in this way. Are they thumbing their collective noses at the U.S. Senate or have they indeed perfected EPT and are using it covertly, and don't want to have to 'fess up about it? Either way, something isn't right.

If they have not actually used HAARP to do EPT, that might be a good thing, after all. In *Planet Earth, The Latest Weapon of War* Dr. Bertell wrote:

> To complete the military investigation of the whole Earth system it was necessary to probe the solid Earth itself, and this again involves the use of wave technology.
>
> Ionospheric heaters such as HAARP create extremely low frequency (ELF) waves which are reflected back to the Earth by the ionosphere. These rays can be directed though the Earth in a method

called deep earth tomography. Since the beamed radiation used to convert the direct electrical current of the electrojet into alternating current must be pulsed, it is reasonable to assume that the ELF radiation it generates will also be pulsed. Pulsed ELF waves can be used to convey mechanical effects, vibrations, at great distances through the Earth. By studying their 'shadow' — that is, where the vibrations are interrupted — it is possible to understand and reconstruct the dimensions of underground structures.

The 10-Hertz ELF wave can easily pass through people, and there is concern that since it corresponds to brain wave frequency it can disrupt human thought. ... However, such waves may also have a profound effect on migration patterns of fish and wild animals as they rely on undisturbed energy fields to find their way. Moreover, the wider effects of deep-earth tomography are unknown. Certainly, it has the capability to cause disturbance of volcanoes and tectonic plates, which in turn have an effect on the weather. Earthquakes, for example, are known to interact with the ionosphere.

Deep Earth probes appear to be an integral part of the military's aim to control and manipulate natural Earth processes. Whilst the potential of ELF waves to generate Earth movements, with associated freak weather, is frightening enough, it is also clear that the interaction between the earth and ionosphere that takes place during ELF generation and transmission may be capable of inducing more direct weather effects.

This heating also turns the excited region of the ionosphere into a plasma (an electrically charged gas) that goes out into space, creating a plume theoretically capable of destroying anything electronic (like an ICBM or a spy satellite) that passes through it.

"… for DOD purposes." And what are DoD purposes? Why, to win wars! DoD related research then would be to make new weapons or to make existing weapons and personnel more effective (which as we learned in the last chapter is called a *force multiplier*).

So, the statement that "The heart of the program will be the development of a unique ionospheric heating capability to conduct the pioneering experiments required to adequately assess the potential for exploiting ionospheric enhancement technology for DOD purposes" means, simply, that the DoD wants to know if they can use this technology to turn the atmosphere into a weapon, or use it to improve existing weapons — this is called <u>weaponizing</u>.

As I have been asking for almost a decade now, is weaponizing the atmosphere a good idea? What about unintended and unexpected consequences? What happens when the enemy responds in kind? And why is it that the mainstream press is so silent on these issues?

In the decade since HAARP first piqued the world's curiosity, interest in HAARP has only grown. Rumors continue to circulate about what HAARP is being used for and why.

Some claims about HAARP are clearly nonsense. How, for example, could HAARP have been used in the first Gulf War, as some have claimed, if it hadn't even been built yet? On the other hand, ridiculous sounding claims may be valid. The problem is trying to tell the difference.

What about claims that HAARP can, or may someday, control the weather? Dr. Begich and Ms. Manning covered this extensively in *Angels*, as did I in *HAARP*. Every year the Internet has been flooded with speculations about HAARP and the weather. Some of the stories have claimed that HAARP was used to move hurricanes, even to dissipating them. The basis of all this is in the original APTI patents, the first of which was granted to Dr. Bernard Eastlund for a "Method And Apparatus For Altering A Region In The

Earth's Atmosphere, Ionosphere, And/Or Magnetosphere." It described several methods for using this apparatus to manipulate the weather. But Dr. Eastlund's patent was for an antenna field forty miles on a side! HAARP, at about a thousand feet on a side, has only about 3% of the output strength Eastlund calculated as necessary to affect the weather. There is nothing but circumstantial and hearsay evidence to link HAARP to any particular weather event — of course, men have been hung with less...

Other stories claim that Enron and other oil and energy firms use HAARP to drive down winter temperatures and drive up summer heat, increasing profits for companies that sell heating oil and electricity to run air conditioning.

Conrad C. Lautenbacher, U.S. Under-Secretary for Commerce, was quoted in the *Wall Street Journal* on 28 July 2003, saying "40% of the $10 trillion U. S. economy is affected by weather and climate changes." Certainly, if you could control the weather, even a little, you would have a tremendous impact on the U. S. economy. A one- or two-degree temperature change could mean millions of dollars made or lost to energy suppliers. Enron was heavily invested (to the tune of a billion dollars!) in "weather derivatives" — literally betting, via the stock market, if the temps would go up or down on future dates. But could Ken Ley, Enron's CEO, really have had a "hot line" to HAARP?

The evidence to support or refute these claims is not only incredibly technical, but in some cases involves "sciences" that don't even exist in the West! Understanding HAARP may involve understanding "Scalar Electromagnetics" or "Harmonic Systems Entanglement" or some other arcane and possibly questionable "alternative" physics. If that sounds over your head, you are not alone!

I am not a scientist, nor a technician — I am a writer, a researcher, and occasionally an activist. I don't know how to build weapons, only how to put sentences and paragraphs together. While HAARP is an incredibly technical subject, mine are not technical books (you may have noticed). I write about the potential impact of this technology on our lives — on you — but not so much about how this stuff actually works, because frankly, I don't have a clue.

I have spent my life reading and thinking about technology and how it has changed our society. As a youth I read voraciously, from wild science fiction novels to the ancient classics, to American history to global politics. But the books I really liked, and read over and over, were the ones that dealt with how societies came to grips with, or were destroyed by their science. My books are about how technology directed to achieving military goals and possibly covert political agendas might affect us. They are also an open invitation to you to research this material for yourself. I don't want you to believe me. I want you to do your own "due diligence."

If you want to delve into "suppressed" sciences try reading books like *Fer de Lance: A Briefing On Soviet Scalar Electromagnetic Weapons* by Lt. Col Thomas E. Bearden (ret). If the Russians are waging a covert war against us, as he claims, this is where you'll get the technology explained.

On the other hand, some of the "wild theories" of mainstream science have, in recent years, been "proven" in the lab. In Chapter Two of *HAARP: The Ultimate Weapon of the Conspiracy* I wrote about Nikola Tesla, the "mad genius" who developed alternating current and was the real discoverer of radio. He gave us the Tesla Coil (without which television is impossible) while trying to develop the wireless transmission of electrical energy. I put forward the hypothesis that he used his world broadcast facility on Long Island to cause a mysterious explosion in Siberia in 1908.

Mainstream scientists are generally agreed that the Tunguska Event as it is known, was caused by a piece of a comet exploding in the atmosphere. I related how Dr. Andrija Puharich hypothesized that the Tunguska Event could have been created through the use of "quantum pairs." The week after I sent the manuscript for that book off to the publisher it was reported around the world that scientists at CERN (the European Organization for Nuclear Research) had successfully "teleported" a photon, using a variant on quantum pair technology!

CERN is the world's largest particle physics center, a place where physicists go to explore what matter is made of and what forces hold it together. The CERN experiments, conducted in Austria and Italy, pioneered "teleported" information about photon polarization (not the photon itself) from a sender to a receiver using the trick of "entanglement"—a deep quantum mechanical connection between particles that was first pointed out by Einstein, Podolsky and Rosen—and described by Puharich as a way that Tesla might have produced the Tunguska Event! In 2002 a team of researchers from the Australian National University replicated the experiment, carrying out a successful teleportation experiment in a gravitational wave lab in Canberra.

This subject of "entanglement" comes up again in the works of Joseph P. Farrell in his attempts to show that the Great Pyramid at Giza was in fact a weapon of mass destruction. The connection to HAARP is seen where Farrell writes, on page 238 of *The Giza Death Star*:

> Eastlund's "HAARP" patents further corroborate the idea that the electromagnetic hydrodynamic properties of the Earth's atmosphere and magnetosphere can be weaponized by the same basic technological and scientific principles for a variety of defensive and offensive purposes.

Dr. Bernard J. Eastlund's name comes up a lot in researching "off trail" science. In 1970 a proposal to develop a "fusion torch" was presented at a New York aerospace science meeting by Drs. Eastlund and William C. Cough. The basic idea was to generate a fantastic amount of heat—a least fifty million degrees Celsius—that could be contained and controlled. Such a torch could be used for a number of purposes, such as tunneling and mining underground.

HAARP TIMELINE

Dr. Bernard J. Eastlund received his patent for a "Method And Apparatus For Altering A Region In The Earth's Atmosphere, Ionosphere, And/Or Magnetosphere" (U.S. Patent Number 4,686,605) on 11 August 1987 while working for the ARCO subsidiary APTI. It was but the first of a dozen related patents that scientists on APTI's payroll would take out over the next few years. These would form the intellectual property that the HAARP program is based on—or not, depending on who you listen to.

Officially HAARP was conceived over two years later, on the morning of 13 December 1989, when a joint Navy-Air Force meeting was held at the Office of Naval Research (ONR) in Washington, D.C. It has since been described as a discussion of their mutual interest in carrying out a DoD program in the area of ionospheric modification. Military and HAARP documents insist that it was at this meeting that the need for a unique heating facility to conduct "critical experiments" relating to potential DoD applications was identified.

The official tale of the birth of HAARP claims that the Navy and Air Force personnel at that after-breakfast meeting at ONR decided to bring the Defense Advanced Research Projects Agency (DARPA) in on the project. Consequently, Navy and Air Force personnel trooped over to DARPA later that day to present their proposal for a DoD sponsored program. As well as representatives of DARPA, people from the Office of the Defense Director of Research & Engineering (DDR&E) were also present at that second meeting of the day.

Another joint Navy-Air Force meeting was held at the Geophysics Laboratory at Hanscom AFB on 24 January 1990, to develop a plan to achieve the "emerging DoD objectives." Another meeting with DDR&E at ONR was held on 12

February 1990, to present the HAARP plan, and to discuss its implementation.

To again quote from the *Executive Summary* of "Applications and Research Opportunities Using HAARP:"

The HAARP facility, currently under development in Alaska, is the outgrowth of more than 30 years 1964 of ionospheric heating research. A wealth of experimental studies conducted at ionospheric heating installations, such as the ones in Arecibo, Puerto Rico; Tromso, Norway; Fairbanks, Alaska; and several installations in the former USSR brought the understanding of the physics and the phenomenology of the HF-ionosphere interactions to a new plateau. The scientific field was ready to make the transition from pure research to applications in the civilian and military arenas. In February 1990, a major workshop sponsored by the Office of Naval Research and Air Force's Phillips Laboratory, and with representations from the National Science Foundation, took place in New London, Connecticut. The workshop, attended by more than 60 representatives from key science, technology, and application areas, defined the operational requirements of the next HF ionospheric transmitter and presented the rationale that led to the HAARP. It was concluded that an HF transmitter located in the auroral zone, with ground power three times larger than the one in Tromso and operationally enhanced with the flexibility provided by the most advanced phased array and software technology, would provide the Nation with unprecedented capability to locally control the state of the ionosphere. The workshop endorsed the HAARP transmitter as the cornerstone of the transition from ionospheric research to technology and applications.

Note that last bit, that HAARP would "provide the

176

Nation with unprecedented capability to locally control the state of the ionosphere." Wow! Being able to control the ionosphere would allow "the Nation" to do a lot of things, from controlling radio communications to the weather (at least "locally").

In short order the project moved off of the drawing boards and into reality. The Appropriation Act for Fiscal Year (FY) 1990 provided funds for the creation of HAARP, jointly managed by the Air Force Research Laboratory and the Office of Naval Research. Three contracts were awarded to ARCO Power Technologies (APTI) to begin feasibility studies in FY1991. In FY1992 the principal contract to begin construction was awarded, also to APTI.

At the same time the environmental impact process began. The MITRE Corporation produced the "Environmental Impact Analysis Process #1," the draft environmental impact study proposal for HAARP, in February of 1993. Later, they also produced the second volume of the Final Environmental Impact Statement, during July of 1993; and the "Electromagnetic Interference Impact of the Proposed Emitters for the High Frequency Active Auroral Research Program (HAARP)" report, of 14 May 1993.

The Environmental Impact Statement was filed with the Environmental Protection Agency and made available to the public by *Federal Register* announcement 23 July 1993. That statement was approved later that year when James F. Boatright, Deputy Assistant Secretary of the Air Force (Installations) published the "Record of Decision (ROD), Final Environmental Impact Statement (FEIS)" on 18 October 1993.

In early November 1993 the United States Air Force announced, via press release, that the prime contractor on the HAARP program was Arco Power Technologies, Incorporated, (APTI). Much later, anti-HAARP investigators discovered that APTI was listed in a Dun & Bradstreet publication (*America's Corporate Families*, 1993, Volume I, page 156) as having a President in Los Angeles, California, and a CEO and a staff of 25 in Washington, D.C. It was cited as having $5,000,000.00 in sales a year.

Dr. Nick Begich points out that the HAARP contract was for an amount five times that company's annual sales. He also notes that they were granted several exemptions to the usual military procurement process. This he believed was evidence that APTI possessed proprietary information requiring that they alone were capable of carrying the project forward, i.e., the Eastlund patents. Since the beginning of the project the patents and the company that owned them have changed hands, and each time the new owner got the contract. This is important to understanding the government's deception in this affair, as at all levels of HAARP, the government and the university scientists connected with the project deny any connection between Dr. Eastlund's Star Wars project and their peaceful little scientific experiment. Note also a contradiction between a broadly published Air Force/Navy fact sheet which claimed that APTI had been granted the contract for HAARP as the result of "a competitive procurement process" and the fact that they were actually granted special privileges and exemptions.

Initial prototype construction began at the Gakona, Alaska, site in late 1993 and was completed a year later in late '94. During that time ARCO divested itself of APTI selling the firm to a major defense contractor, E-Systems of Dallas, Texas.

About the time that APTI was sold to E-Systems Dr. Eastlund formed his own company, Eastlund Scientific Enterprises Corporation (ESEC). Although the folks at HAARP vehemently deny any use of their facility for weather modification purposes, Eastlund announced in 2002 that ESEC had completed a contract with the European Space Agency to review the weather modification potential of the HAARP facility! ESEC has also contracted to perform numerical simulations of tornado suppression with high power electromagnetic radiation produced with Solar Power Satellites. Two papers (available on ESEC's website) have been published on this, thus far.

In 1994 the United States Senate froze funding for HAARP until planners increased emphasis on earth penetrating tomography (EPT) uses for nuclear counterproliferation efforts.

In a Program Research & Development Announcement (PRDA) with the title "Detection and Imaging of Underground Structures Using ELF/VLF Radio Waves" the Space Effects Division of the Phillips Laboratory Geophysics Directorate asked for research proposals related to the "theoretical understanding and practical development and demonstration of techniques for the detection of underground structures using ELF/VLF radio waves generated by natural and man-made sources." From it we can get an idea what the funding freeze at HAARP was all about. In that PRDA they wrote:

> ELF/VLF radio waves penetrate deeply beneath the surface of the earth and interact with the geologic structure of the earth. This interaction induces secondary fields with measurable effect at and above the surface of the earth. Proper understanding of the physics of the generation and propagation of ELF/VLF waves and their interactions with earth materials will allow these waves to be used for applications such as sub-surface communications and exploration of the subsurface geological structure. The research called for under this effort is to assess the viability of exploiting the concept of electromagnetic induction to detect and image subterranean features such as tunnels, bunkers, and other potential military targets.
>
> Geophysical surveying using natural ELF/VLF sources, such as lightning or auroral generated signals, is an established procedure. In general, however, the procedure has been developed with the interest in locating areas of highly conductive material such as

179

metal ore deposits.

In 1994, the Senate Armed Services Committee noted the promising results of the high frequency active auroral research program (HAARP). This transmitter in Alaska, besides providing a world class research facility for ionospheric physics, could allow earth- penetrating tomography over most of the northern hemisphere. Such a capability would permit the detection and precise location of tunnels, shelters, and other underground shelters. The absence of such a capability has been noted as a serious weakness in the Department of Defense plans for precision attacks on hardened targets and for counterproliferation. In fact, the May 1994 report from the Deputy Secretary of Defense on nonproliferation and counterproliferation activities and programs recommends increased funding of $75.0 million annually for detection of underground structures. The committee recommended $5.0 million in PE 62601F to continue the HAARP project, but notes with concern that the capital cost of a full-scale HAARP facility could be as much as $90.0 million. Unless the Department of Defense is committed to include such a project in future budget requests, the recommended authorization for fiscal year 1995 will have little effect. Therefore the committee directed that none of these funds may be obligated or expended until the Secretary of Defense notifies the Committees on Armed Services of the Senate and the House of Representatives that the Department will, as part of the nonproliferation and counterproliferation program recommended in the May 1994 report, include the cost for a full-scale HAARP facility in its fiscal year 1996 budget request.

About this same time the above mentioned scientific committee was sponsored by the Air Force's Phillips

Laboratory and the Office of Naval Research and convened by the East-West Space Science Center of the University of Maryland, of which Roald Zinurovich Sagdeev was the Director. The committee was composed of:

- Dennis Papadopoulos, Committee Chairman; Professor of Physics, University of Maryland
- Paul A. Bernhardt, Active Experiments Project Leader, Beam Physics Branch, Plasma Physics Division, Naval Research Laboratory
- Herbert C. Carlson, Jr., Deputy Chief Scientist, Geophysics Directorate, Phillips Laboratory
- William E. Gordon, professor and former Dean, Rice University; Member of the National Academy of Sciences
- Alexander V. Gurevich, Head, Ionospheric Division, Lebedev Institute; Corresponding Member of the Russian Academy of Sciences
- Michael C. Kelly, Professor of Electrical Engineering, Cornell University
- Michael J. Keskinen, Beam Physics Branch, Plasma Physics Division, Naval Research Laboratory
- Roald Z. Sagdeev, Distinguished Professor of Physics, University of Maryland; Member of the Russian Academy of Sciences; Foreign Member of the National Academy of Sciences
- Gennady M. Milikh, Committee Secretary; Research Associate, University of Maryland

This committee released the "Applications and Research Opportunities Using HAARP" report, and under separate cover published a written description of HAARP plans and objectives under the title *Executive Summary* (quoted p.176 from above), which was circulated to Navy, Air Force, and DARPA personnel for coordination.

As I mentioned in the Introduction to this book, the UCLA Plasma Physics Laboratory operates an ionospheric

181

heater near Fairbanks, Alaska, known as HIPAS (for HIgh Power Auroral Stimulation). Dennis Papadopoulos, this committee's Chairman, was one of the principal scientists at HIPAS. It was his work there that proved that ionospheric heaters (in this case HIPAS) could generate ELF to successfully find underground targets.

Dennis Papadopoulos was erroneously identified as the chief scientist on the HAARP project by the BBC/A&E television program *Masters of the Ionosphere*. On it he spoke enthusiastically of the first test of earth penetrating tomography capability. He was talking about HIPAS but the show made it seem like he was talking about HAARP. He said: "There was an experiment in which there was a particular old mine which was about 30 meters underground and we were trying to see whether we could really image it..." using ionospherically generated ELF. According to Dr. Papadopoulos, it worked. Sensors in Fairbanks picked up the ELF waves as they returned from deep underground. After interpretation, they clearly revealed the old mining tunnel below the surface. Dr. Papadopoulos told the camera:

> This was really the first test of the concept in the world and we were, I was actually amazed because usually a first experiment always fails. This experiment succeeded beyond our imagination. It is a fantastic remote sensing tool. I can do remote sensing of the ionosphere, of the ground, of the underground, of the seas. I can measure the temperature of the water. I think, you know, dreams that I have dreamt for the last, you know, 25 years are becoming reality and the first one happened two weeks ago. That was really a sweet moment.

Note also the three Russians involved in drafting the HAARP plan via this committee. Before coming to the U.S. Dr. Sagdeev had been the Director of the Soviet Union's Space Research Institute (comparable to our National Aeronautics and Space Administration (NASA)) for fifteen years. While there he directed many high-profile multinational projects, including the joint U.S.-U.S.S.R. Apollo-Soyuz program, and international missions to probe Halley's comet and later to Phobos, a moon of Mars. These last two projects were devised and implemented by Academician Sagdeev in cooperation with more than twelve countries.

Before his appointment to the Space Research Institute in 1973, he had an impressive career in nuclear science as a plasma physicist; so much so that he was one of the youngest scientists ever elected to full Academician of the U.S.S.R. Academy of Sciences. At the time I wrote *HAARP* he was listed as Director Emeritus, Russian Space Science Institute, and was in the United States as a Distinguished Professor of Physics at the University of Maryland. He was also a Foreign Member of the United States National Academy of Sciences and a Senior Associate at the Center for Post-Soviet Studies.

Shortly after this committee issued its report, in 1995, Academician Sagdeev was co-recipient, with Evgany P. Velikhov, of the Leo Szilard Award for Physics in the Public Interest "for their unique contributions to Soviet Glasnost which was a major factor in reversing the nuclear arms race between the Soviet Union and the United States." This award is presented annually by the American Physical Society "to recognize outstanding accomplishments by physicists in promoting the use of physics for the benefit of society in such areas as the environment, arms control, and science policy." Leo Szilard (1898 - 1945) is credited with being the physicist who first conceived of the idea of building an atomic bomb back in the 1930s. At his urging Albert Einstein (1879 - 1955) wrote a letter to President Franklin D. Roosevelt in 1940. That led to the creation of the Manhattan Project that eventually gave us The Bomb.

Velikhov and Sagdeev received the Leo Szilard Award in recognition of their having organized the Soviet Scientists' Committee for Peace Against the Nuclear Threat, in March of 1983. This committee published a physics-based critique of the Strategic Defense Initiative (SDI) three years later. In Professor Sagdeev we see a very interesting connection between HAARP, SDI, plasma physics (HAARP is alleged to be intended to study the plasma physics of the upper atmosphere), the environment, and arms control; subjects that came up repeatedly in my search for an understanding of what HAARP is all about.

Why would a former top Soviet official, scientist, and SDI expert be a part of the birthing of HAARP? I believe that HAARP is SDI technology coming on-line. When President Reagan announced SDI he promised to share the technology with the Soviets. Some thought he was wacky for saying that, others applauded the gesture. Was the presence of three Russians on that committee proof that HAARP is "Star Wars" and that we really were planning to share it with them?

On 3 April 1995 the *Wall Street Journal* announced "Raytheon to Acquire E-Systems for $64 a Share..." It would seem that selling HAARP to Raytheon was what it took to get the project moving again. A press release dated 4 December 1995 under the heading "New Defense Law Contains Alaska Projects" reads:

> The $243 billion defense appropriations bill that became law late last week contains several Alaska-specific items. At the request of Senator Stevens, chairman of the Defense Appropriations Subcommittee, the bill continues the local hire provision for Department of Defense service and construction projects in Alaska. The Alaska joint military exercise, Northern Edge, will receive $5 million, and $15 million is included to continue the High Altitude [sic] Auroral Research Program (HAARP), at Stevens' request.

Senator Ted Stevens, Nick Begich said, hyped HAARP in

his home state during the 1995-6 election year as some great Godsend. When emotionally defending HAARP before his committee, Stevens said:

> I could tell you about the time when the University of Alaska came to me and said it might be possible to bring the aurora to Earth. We might be able to harness the energy in the aurora... No one in the Department of Defense, no one in the Department of Energy, no one in the executive branch was interested in pursuing it at all. Why? Because it did not come from the good old boy network. So I did just what you say I should do. I got Congress to earmark the money, and the experiment is going on now. It will cost $10 million to $20 million. If it is successful, it will change the history of the world.

Frankly, I am mystified by his statement. Nowhere in the HAARP documentation, official or otherwise, is there any indication that HAARP originated with the University of Alaska or Senator Stevens. More evidence of conspiracy and cover-up?

Earlier that year, 1995, Raytheon bought E-Systems and all the APTI patents they held, and, perhaps thanks to Senator Stevens, Congress budgeted $10 million for HAARP for 1996 under "Counterproliferation—Advanced Development" spending. In the 1997 Descriptive Summary of the Counterproliferation Advanced Development Budget HAARP appears under the sub-heading "Project P539 Counterforce." There it is recorded that "In FY96, Congress added $10 million to be used for the High-Frequency Acoustic [sic] Auroral Research Program (HAARP) to this project." Elsewhere in that report it states "... in FY96 only, the Congressionally added HAARP program funds will be used to explore the ability of auroral transmissions to detect and locate underground structures of the type where WMD can be developed or stored."

The HAARP final ionospheric research instrument (FIRI) was planned to be a field of 180 antennas arranged in a rectangular grid of 12 rows by 15 columns. Initially a smaller set of elements was constructed so that the predicted performance could be verified before the entire facility was built. That initial phase of the program was called the Developmental Prototype (DP). By April of 1995 the DP array of 48 antenna towers arranged as 8 columns by 6 rows had been completed. Thirty additional unpowered and unused towers were also erected at that time.

The first round of tests of the DP was in April 1995. More start-up tests were conducted in July and November of 1995, while tests of the aircraft detection radar were conducted in September of that year. The aircraft alert radar (AAR) is intended to automatically shut off "appropriate transmissions" when aircraft are detected within, or approaching a "safety zone" established around the HAARP site.

HAARP documents claim that the facility was shut down at the end of the last set of initial low power tests on the DP on 21 November, 1995. Officially, no testing was conducted from that time until the HAARP facility was at last put to scientific use for the first time, over a year later. A two week flurry of scientific research activity, called a "campaign," took place from 27 February to 14 March 1997. In addition to science experiments, this two week period included several visits from tour groups; participation in a lecture series by HAARP personnel at the nearby community college; a public talk on ionospheric research and the HAARP facility; and the first HAARP-Amateur radio listening test in which ham radio enthusiasts where invited to tune in for the HAARP signal.

During the early part of the August 1997 testing period several experiments were performed with the NASA WIND satellite which was at a favorable position in its orbit. The third annual HAARP Open House was held 23-24 August, 1997. Program personnel were present to discuss the project

186

and to give demonstrations and tours of the facility. Several experts in ionospheric physics were also present to discuss the research plans and the physics of the earth's upper atmosphere.

By this time Nick Begich and Jeanne Manning's book had been on the market for two years and Nick had appeared on the radio hundreds of times and managed to get the issue of HAARP heard by important and well-placed people. As Michel Chossudovsky reported in his article "It's Not Only Greenhouse Gas Emissions: Washington's New World Order Weapons Have The Ability To Trigger Climate Change," Nick succeeded in bringing HAARP before the European Parliament. Professor Chossudovsky wrote:

> In February 1998, responding to a report of Mrs. Maj Britt Theorin — Swedish MEP (Member of the European Parliament) and longtime peace advocate — the European Parliament's Committee on Foreign Affairs, Security and Defence Policy held public hearings in Brussels on HAARP. The Committee's 'Motion for Resolution' submitted to the European Parliament:
>
> Considers HAARP... by virtue of its far-reaching impact on the environment to be a global concern and calls for its legal, ecological and ethical implications to be examined by an international independent body...; [the Committee] regrets the repeated refusal of the United States Administration... to give evidence to the public hearing ...into the environmental and public risks [of] the HAARP program.
>
> The Committee's request to draw up a 'Green Paper' on 'the environmental impacts of military activities', however, was casually dismissed on the grounds that the European Commission lacks the required jurisdiction to delve into 'the links between environment and defense'. Brussels was

187

anxious to avoid a showdown with Washington.

To recap… After APTI got the initial contract they erected the first few towers and wired them up to see if the various bits of gear they had purchased would work together. They did, so they put up a bunch more towers and wired some of them and ran some tests to see if the field would out-put radio waves the way they thought it would. That worked too, so they started running experiments in two-week long sessions called campaigns.

Everything went swimmingly on site, but things got hot in the world's press. *Mother Jones Magazine* cited HAARP as one of the Top Censored News Stories of 1994. Similarly Project Censor listed it as one of the top underreported stories of the year and again listed it as one of the 100 most underreported news stories of the 20th century in their 2000 Millennial Edition. Even worse for the folks at HAARP the prestigious international publication *Jane's Defense Weekly* listed HAARP as a weapons system!

Things went from bad to worse for the folks at HAARP after the turn of the century. They wanted to finish erecting the final ionospheric research instrument (FIRI) by 2002 but got their budget slashed instead. It appears that the incoming President, George W. Bush, was more concerned with his anti-missile missile defense program than with ionospheric research, or exotic electromagnetic weapons research, depending on which HAARP really is, and killed HAARP's funding for the first two years of his administration.

Adding insult to injury in August of 2002 Russia's Interfax news agency reported that The Russian State Duma had expressed concern about HAARP, calling it a program to develop "a qualitatively new type of weapon." An appeal, signed by 90 deputies, demanding that an international ban be put on such large-scale geophysical experiments was sent to President Vladimir Putin, to the United Nations and other international organizations, to the parliaments and leaders of the UN member countries, to the scientific public

188

and to mass media outlets.

After 2000 there were few changes at the HAARP site. The biggest change was that they found one of the two large diesel generators that were supposed to have gone into the power plant. That is the large white building seen in the aerial photographs. It was built prior to the HAARP project, when the site was an over-the-horizon radar facility. The site had been abandoned when only partially constructed. The power plant building had been erected but the generators never got installed. HAARP got that one generator installed and moved the offices and some of the control facilities out of temporary modular buildings into the power plant. Before the big generator was installed, the IRI was powered by many small generators located in the shelters with the transmitters. I have heard stories of scientists sitting at their desks at HAARP drumming their fingers and grousing that there was nothing to do because there wasn't enough money to fill the fuel tanks on the generators!

As of January 2002 the working portion of the IRI consisted of 48 antenna elements arranged as a rectangular array of 8 columns by 6 rows. The total power capability of this transmitter system was 960 kilowatts. It looks like the Final IRI (FIRI) with an effective radiated power of 3.6 megawatts was finally completed late in 2005 or early in 2006, as I will detail in a moment.

Soon after the project began the US Navy's Office of Naval Research (ONR) put up a website for HAARP, and the Air Force also put up an informational webpage. Later the University of Alaska put up a site, mirroring the Navy site. The Navy website has since been taken down. Unfortunately, the University of Alaska's HAARP site is very poorly maintained. With the exceptions of a few small entries the site has not been updated for over three years. Another way to keep secrets? The most up-to-date information on HAARP comes from the Defense Advanced Research Projects Agency's site, as we will see shortly.

Initially HAARP was jointly managed by ONR and the

189

Air Force's Phillips Laboratory in Massachusetts. In 2002 Project Management of HAARP was transferred to Defense Advanced Research Projects Agency (DARPA). It would seem that DARPA was brought in to whip the project into shape. According to the official DARPA webpage for HAARP their portion of the project—completing the FIRI—is now done and HAARP is being "transitioned" back to the Air Force and Navy in FY2006.

COMPLETING HAARP

Back when HAARP was first under construction in the early 1990s a chain link fence was erected around it. The University hired a security guard to roam the grounds. A very large sign was placed by the front gate declaring it to be the property of the Department of Defense and jointly operated by the Air Force and Navy with the logos of all these agencies on it. After *Angels* was published someone stole the sign. That made the University mad, so they fired the "rent-a-cop." Then someone stole the gate! Last I heard the property was wide open and if you could get there you could just drive in and talk to anybody who was around. That was before the role of Project Manager was given to a research scientist at DARPA, so things may be different there now.

What is DARPA? If you have seen any of the James Bond (*007*) movies you may remember *Q*, the guy whose research team came up with all those crazy gadgets for Bond. Now imagine Q's department being a vast government bureaucracy with a budget in the millions (billions?) and workers in the thousands (tens of thousands?)—that's DARPA!

Per their official fact sheet:

The Defense Advanced Research Projects Agency (DARPA) is the central research and development organization for the Department of Defense (DoD).

190

It manages and directs selected basic and applied research and development projects for DoD, and pursues research and technology where risk and payoff are both very high and where success may provide dramatic advances for traditional military roles and missions.-

Note that "high risk" element. Here we may be talking about several kinds of risk including political and environmental. Back before the break-up of the Soviet Union, when HAARP began, many were concerned that the Star Wars weapons program would upset the global balance in that it might give the US "first strike" capability and at the same time negate Russia's ability to retaliate, thus destroying Mutual Assured Destruction (MAD), the foundation of Cold War détente — possibly leading to a preemptive Russian first strike and the outbreak of World War III. On the other hand, some researchers suggested that HAARP was capable of severe disruptions of natural systems, from weather out of control to initiating a magnetic pole shift or reversal, to other bits of environmental mayhem too horrible to contemplate — up to and including destroying all life on earth! Talk about high risk!

There are eight technical offices in DARPA. HAARP was placed under the Tactical Technology Office (TTO). According to their fact sheet:

> The Tactical Technology Office engages in high-risk, high-payoff advanced technology development of military systems, emphasizing the "system" and "subsystem" approach to the development of Unmanned Systems, Space Systems and Tactical Multipliers.

Note that "tactical multipliers" — this may be a tip-off that HAARP was seen by DARPA as some sort of force multiplier, possibly engaged in ionospheric modification as outlined under *Weather as a Force Multiplier: Owning the Weather in 2025* in the previous chapter.

DARPA signed a Memorandum of Agreement (MOA) with the Air Force and Navy to run this program for them in November 2002.

On 17 February 2003 a press release went out announcing that BAE Systems North America had reached a definitive agreement with Advanced Power Technologies, Inc. (APTI) to purchase the corporation for $27 million in cash (note that somewhere along the line the "A" in APTI changed from "ARCO" to "Advanced"). That press release has since been deleted from BAE's website. Details on this purchase have completely vanished from the Internet. I called a spokesperson at APTI and got very little information. Indeed, when I asked about HAARP the poor fellow had never heard of it — or so he claimed!

One year and two months after purchasing APTI, BAE Systems announced that they had received a contract from the Navy for $35 million to complete the HAARP FIRI. After purchasing APTI, BAE Systems became the owners of the intellectual property, the patents, that make HAARP possible. When HAARP's funding was resumed they automatically got the contract. This is important evidence that HAARP is a ground-based weapons system, as laid out in those patents. In a press release dated 10 June 2004 BAE Systems proclaimed:

> The Office of Naval Research has awarded BAE Systems a $35.4 million contract to manufacture 132 high frequency (HF) transmitters for installation in the High Frequency Active Auroral Research Program's (HAARP) phased array antenna system. The contract was finalized April 19 with BAE Systems Information & Electronic Warfare Systems in Washington, D. C.
>
> The HAARP program collects and assesses data to advance knowledge of the physical and electrical properties of the Earth's ionosphere. "We look

forward to contributing to this critical program. This is an opportunity for BAE Systems to play an important role in expanding knowledge of the Earth's ionosphere. Significant potential applications include long-range communication, sensing and satellite vulnerability to nuclear effects," said Ramy Shanny, BAE Systems vice president and general manager for Advanced Technologies (AT).

In 1992, AT was awarded a contract to design and build the Ionospheric Research Instrument (IRI), the HAARP program's primary tool used to study ionospheric physics. The IRI is currently composed of 48 antenna elements and has a power capacity of 960,000 watts. When installed, the additional 132 transmitters will give HAARP a 3.6 mega-watt capacity. The HAARP build-out is jointly funded by the U. S. Air Force, the U. S. Navy and the Defense Advanced Research Projects Agency (DARPA).

Although the above-mentioned press release has disappeared from BAE's website, one can still find the government's version online which states:

BAE Systems Advanced Technologies Inc., Washington, D.C., is being awarded a $35,351,790 firm-fixed-fee delivery order under previously awarded indefinite-delivery/indefinite quantity contract (N00014-02-D-0479) to manufacture high frequency transmitters for installation in the High Frequency Active Auroral Research Program (HAARP) Gakoma Facility phased array antenna system. The transmitters will be tested for proper performance in accordance with contract specifications and shipped to the HAARP facility. Work will be performed in Dallas, Texas (72.4 percent) and Washington, D.C. (27.6 percent) and, is expected to be completed in June 2007. Contract funds will not expire at the end of the current fiscal year. The Office of Naval

Research, Washington, D.C., is the contracting activity.

So, what is BAE Systems? The contract was awarded to BAE Systems Information & Electronic Warfare Systems (IEWS) division. Just the sound of "Information & Electronic Warfare" should have been ringing alarm bells—and perhaps it did, as that division has since changed its name to the "kinder and gentler" sounding Electronics and Integrated Solutions Operating Group, which is defined by the company as "a major defense electronics component of BAE Systems, Inc." which is a US corporation and a wholly owned subsidiary of BAE Systems PLC, a multinational corporation headquartered in the United Kingdom.

BAE Systems PLC is the end result of over one hundred years of mergers and acquisitions, with the former British Aerospace at its core. BAE Systems PLC is engaged in the "development, delivery, and support of advanced defense and aerospace systems in the air, on land, at sea, and in space. The company designs, manufactures, and supports military aircraft, surface ships, submarines, radar, avionics, communications, electronics, and guided weapon systems." They have major operations across five continents and customers in some 130 countries, employing more than 90,000 people worldwide, generating annual sales of more than $20 billion through its wholly owned and joint-venture operations.

Per a BAE Systems North America fact sheet:

> BAE Systems North America has grown to become one of the top 10 suppliers to the US Department of Defense—dedicated to solving our customer's needs with both highly innovative and leading edge solutions across the defense electronics, systems, information technology and services arenas.

BAE Systems lost no time in farming out the HAARP contract. Five days after getting the job from the Navy they

194

sub-contracted the transmitter portion of the project to DRS Technologies.

DRS TECHNOLOGIES RECEIVES $23.3 MILLION CONTRACT TO PROVIDE HIGH-FREQUENCY RADIO TRANSMITTERS FOR U.S. GOVERNMENT

Parsippany, NJ, June 15 -- DRS Technologies, Inc. (NYSE: DRS) announced today that it has received a $23.3 million contract, including options, to provide high-frequency (HF) radio transmitters for the High-Frequency Active Auroral Research Program (HAARP), which supports a U. S. government Arctic research facility being built to study the Earth's upper atmosphere.

The $11.5 million base contract was awarded to DRS by BAE Systems PLC (LSE: BA. L). For this award, DRS will manufacture more than 60 Model D616G 10-Kilowatt Dual Transmitters to fulfill the transmitter requirements for the HAARP program. Work for this order will be performed by the company's DRS Broadcast Technology unit in Dallas, Texas. Product deliveries to BAE Systems' Information and Electronic Warfare Systems in Washington, D.C., are scheduled to begin in March 2005 and continue for approximately one year.

"We are pleased to continue our role as a premier supplier of transmitters for the HAARP program," said Steven T. Schorer, president of DRS's C4I Group. "This award enhances DRS's position as a leader in high-technology radio frequency solutions for secure and tactical communications systems supporting the applications of the government scientific research community."

The high-frequency or short-wave Model D616G Transmitters were designed specifically for the U. S. government HAARP research facility. Currently, the ionosphere provides long-range capabilities for commercial ship-to-shore communications, transoceanic aircraft links, and military communications and surveillance systems.

A primary goal of HAARP is to understand how variations in the sun's radiation affect the performance of radio systems and to improve military command, control, communications and surveillance systems.

DRS Broadcast Technology, formerly known as Continental Electronics, is a global leader in broadcast transmitter equipment. It is the foremost supplier of advanced radio frequency transmission technology and the world's most experienced provider of the highest power radio broadcast equipment, offering a full range of products for broadcasting, military and scientific applications.

DRS Technologies, headquartered in Parsippany, New Jersey, provides leading edge products and services to defense, government intelligence and commercial customers. Focused on defense technology, DRS develops and manufactures a broad range of mission critical systems. The company employs 5,800 people worldwide.

Twelve days later the subcontractor who would put up the remaining antennas sent out their own press release:

PHAZAR CORP
P.O. Box 121697 Fort Worth, Texas 76121
(940) 325-3301
NEWS RELEASE
June 27, 2005
ANTENNA PRODUCTS CORPORATION
AWARDED CONTRACT
PHAZAR CORP announced today that Antenna Products Corporation in Mineral Wells, Texas, a wholly owned subsidiary, was recently awarded

196

a $1,733,482 firm fixed price subcontract from BAE Systems for the installation of Low Band and High Band Antenna Matching Unit Assemblies in Alaska. The Low Band and

High Band Antenna Matching Unit Assemblies, complete with cable sets, mounting brackets and various other components, are currently being manufactured by Antenna Products Corporation under a production contract received in February, 2005. The installation work will start in June, 2005 and is scheduled to be completed on November 1, 2005. The equipment will be installed at the High Frequency Active Auroral Research Program (HAARP) ionospheric research site near Gokona, Alaska, the site of the recently completed installation of an array of 132 crossed dipole antennas built and installed by Antenna Products Corporation in 2004.

It looks like they made their November 2005 deadline, or came close. Independent researchers have established that the HAARP site was in near daily use in 2005. As best anyone can tell the technicians from BAE Systems and their subcontractors spent the day installing the new antennas and later wiring them up to the new transmitters, then spent several hours each night broadcasting to check the quality of the workmanship and equipment. As you will see below DARPA announced the completion of this phase of HAARP's history via their website on 9 March 2006.

HAARP TODAY

DARPA's official word on HAARP is as follows:

High Frequency Active Auroral Research Project (HAARP)Program Manager: Dr. Sheldon Z. Meth
OVERVIEW:
The High Frequency Active Auroral Research

Project (HAARP) developed new experimental research capabilities and conducted research programs to exploit emerging ionosphere and radio science technologies related to advanced defense applications. The FY 1990 Appropriation Act provided funds for the creation of HAARP, jointly managed by the Air Force Research Laboratory and the Office of Naval Research to exploit emerging ionosphere and high power radio technology for new military systems applications. Key to the current effort was the expansion of the experimental research facility that includes a 3.6 MW high-frequency transmitter and a variety of diagnostic instruments, to conduct investigations to characterize the physical processes that can be initiated and controlled in the ionosphere and space, via interactions with high power radio waves. Among these were: (1) the generation of

 extremely low frequency/very low frequency radio waves for submarine and other subsurface communication, and the reduction of charged particle populations in the radiation belts to ensure safe spacecraft systems operations; (2) the control of electron density gradients and the refractive properties in selected regions of the ionosphere to create radio wave propagation channels; and (3) the generation of optical and infrared emissions in space to calibrate space sensors. To date, the facility has been developed to include a suite of optical and radio diagnostics and an advanced, modern, high frequency transmitting array that has a radiated power of 960 kW, about one-third of the 3.6MW called for in the original concept and plan. The current high frequency transmitting array has proven to be extremely reliable and flexible, and has shown the feasibility of the overall concept. However, results to date have indicated that the advanced applications-related research activities and new military system concept demonstrations envisioned under the program require that the high frequency transmitting capability at the site be increased from the present 960 kW level to the originally planned 3.6 MW level. A study completed by an Air Force/Navy Panel also pointed to additional high-value functions that can potentially be accomplished with the a 3.6 MW capability, in particular, the exploration and refinement of scientific principles that could lead to the development and deployment of a system to provide protection for spacebased assets from emergent asymmetric threats. DARPA established an MOA with the Air Force and Navy for this program in November 2002. The HAARP technology is transitioning to the Air Force and Navy in FY 2006.

Program Plans

• Completed the HAARP high frequency transmitting array at the HAARP Research Station, Gakona, AK.

• Prepared the existing HAARP facility in preparation for ionospheric testing.

• Conducted advanced ionosphere and radio science research and analysis of applications including space-based asset protection and phenomena related to its implementation.

Updated: 3/9/06

Not only was Dr. Meth in charge of HAARP, he also managed programs called Air Laser and MAgneto Hydrodynamic Explosive Munition (MAHEM)! Does it seem odd to you that a guy running a laser development program and designing some kind of bomb would also be tinkering with how the top of the atmosphere reacts to radio waves?

Notice that throughout this DARPA webpage and the report "Applications and Research Opportunities Using HAARP" they use the term *applications*, as in: "programs to

199

exploit emerging ionosphere and radio science technologies related to advanced defense *applications;*" and "to exploit emerging ionosphere and high power radio technology for new military systems *applications.*" HAARP is not "pure" research. As "Applications and Research Opportunities Using HAARP" put it, it is all about "the transition from ionospheric research to technology and applications" — i.e., *ionospheric enhancement.*

They list four things they've done with HAARP, but only use three numbered bullets:

> (1) the generation of extremely low frequency/ very low frequency radio waves for submarine and other subsurface communication, and the reduction of charged particle populations in the radiation belts to ensure safe spacecraft systems operations;
>
> (2) the control of electron density gradients and the refractive properties in selected regions of the ionosphere to create radio wave propagation channels; and
>
> (3) the generation of optical and infrared emissions in space to calibrate space sensors.

Is it possible that Dr. Meth has trouble counting, or does one way to use the antenna field produce both ELF and reduce the populations of charged particles in the radiation belts?

As repeatedly stated, and admitted to by everyone, although down-played by the civilians at the University of Alaska, one of the prime purposes of the project was to communicate with deeply submerged submarines. This is certainly a valid military use for this technology, although finding a "dual use" civilian side might be a bit of a stretch.

I'll take up the second point, re: charged particles in the radiation belts, in a bit.

Point (2), "the control of electron density gradients and the refractive properties in selected regions of the ionosphere

200

to create radio wave propagation channels" is how HAARP controls radio communications. It has been discovered that injecting energy into the ionosphere literally reshapes it, allowing for the creation of virtual tubes and tunnels in the sky, called wave propagation guides or wave ducting.

Point (3), "the generation of optical and infrared emissions in space to calibrate space sensors," is a bit of a head-scratcher. I can see heating up the atmosphere till it glows, but how do you get "optical emissions" in the void of space?

One of the future uses for HAARP is given as "the exploration and refinement of scientific principles that could lead to the development and deployment of a system to provide protection for spacebased assets from emergent asymmetric threats."

In "Understanding 'Asymmetric' Threats to the United States" by Lambakis, et al., published by the National Institute for Public Policy in September 2002, we read:

> "Asymmetric threats" have come to mean a great many things, and so the term lacks basic definition. Very often, those who use the concept point out that the "asymmetric foe" is one who strives to exploit U.S. weaknesses or who strives to circumvent superior U.S. military power by cunning, surprise, indirect approach, or ruthlessness. … In general, the term is used to describe forms of attack against which the United States has no defenses, and it depicts tactics that Washington will not abide (either because they are morally reprehensible or restricted by legal agreement).

The terrorist attacks on the U.S. on 11 September 2001 have been described as "asymmetrical." (One liberal author quipped on PBS's *Now With Bill Moyer* that they were also a "faith-based initiative.") Some would argue that HAARP is actually not intended to "provide protection for spacebased assets from emergent asymmetric threats" but actually *is* an asymmetric threat to other nations' spacebased assets — and possibly our own! The space shuttle Columbia broke up during re-entry on 1 February 2003 killing all seven astronauts aboard. Some researchers believe HAARP may have played a role in its demise!

SPACE SHUTTLE COLUMBIA

Space Shuttle *Columbia* (NASA Orbiter Vehicle designation: OV-102) was the first space shuttle in NASA's orbiter fleet to enter Earth orbit and was always considered the flagship of the fleet.

Enterprise (OV-101) was the first space shuttle built, but it never flew in space. It was originally to be named *Constitution* in honor of the 200th anniversary of the *U.S.S. Constitution* ("Old Ironsides"). However, viewers of the popular sci-fi TV show *Star Trek* started a write-in campaign urging the White House to name it Enterprise. OV-101 was rolled out of the assembly facility at Rockwell's Air Force Plant 42, Site 1 in Palmdale California on 17 September 1976. Nine years later it was ceremonially retired. On 18 November 1985 it was taken from the Kennedy Space Center to Dulles Airport, Washington, D.C. where it was formally presented to the Smithsonian Institution. The Enterprise was built and used as a test vehicle and was never equipped for space flight.

Following the Enterprise was the orbiter *Columbia*. Its first mission, STS-1, lasted from 12-14 April 1981. Four sister ships joined the fleet over the next 10 years: *Challenger*, arriving in 1982 but destroyed four years later; *Discovery*, 1983; *Atlantis*, 1985; and *Endeavour*, built as a replacement for Challenger in 1991.

Construction began on Columbia in 1975 primarily in Palmdale, California. Columbia was named after the Boston-based sloop *Columbia* captained by American Robert Gray, who explored the Pacific Northwest, and became the first American vessel to circumnavigate the world. The name also honored *Columbia,* the Command Module of Apollo 11.

Dr. Bernard Eastlund wrote three of the twelve patents for a HAARP-like technology that were assigned by his team to their boss APTI. All twelve of the APTI patents are listed in Appendix B. His second patent was for a "Method for producing a shell of relativistic particles at an altitude above the Earth's surface." This patent described a method

> ... for establishing a region of a high density, high energy plasma at an altitude of at least about 1500 kilometers above the Earth's surface. Circularly polarized electromagnetic radiation is transmitted at a first frequency substantially parallel to an Earth's magnetic field line to excite electron cyclotron resonance heating in normally occurring plasma at an altitude of at least about 250 kilometers to generate a mirror force which lifts said plasma to said altitude of at least about 1500 kilometers. Heating is continued at a second frequency to expand the plasma to the apex of said field line whereupon at least some of the plasma is trapped and oscillates between mirror points on said lines. The plasma will be contained within adjacent field lines and will drift to form a shell of relativistic particles around a portion of the Earth.

Simply put, this creates a "death zone" for anything electronic flying through this "shell of relativistic particles." It would fry out the electronics of anything moving through it such as a spy satellite or a nuclear missile. This patent was granted in 1991 as United States Patent 5,038,664.

According to *Defense News,* 13-19 April 1992, the U.S. deployed an electromagnetic pulse weapon (EMPW) in Desert Storm (Gulf War I). It was designed to mimic the flash of electricity from a nuclear bomb, the electromagnetic pulse (EMP). Per *The Language of Nuclear War: An Intelligent*

Citizen's Dictionary:

An EMP is a burst of radiation released immediately after a nuclear explosion. The EMP is essentially an electric field and a magnetic field moving away from the blast. The electromagnetic pulse burns out electronic circuitry, destroying communications systems, computers, and other sophisticated electronic instruments. The implications of the EMP are uncertain, but some experts assert that the EMP released by one large nuclear explosion over the central United States could cause an electrical blackout affecting the entire country. Further, it is possible that the EMP would damage the circuitry in missiles so that they would be unable to reach their targets. The EMP was first detected during a nuclear test at Johnston Island in 1962.

Early HAARP documents discussed how the facility could be used to create artificial EMPs. After Eastlund went public all such mentions disappeared from official sources. Many researchers are convinced that HAARP can cause EMP-like effects in the upper atmosphere. HAARP transmissions excite beta particles or electrons in the magnetosphere making them move very fast, nearly to speed of light—Eastlund called these *relativistic particles*. This is what would damage the electronics of a spacecraft or missile flying through the region being bombarded by the HAARP signal, or the plume of material so created, reaching up from the ionosphere out into space. The shell, cloud, or plume of relativistic particles would essentially mimic the effects of an EMP with the exception that it would be continuous, and so would be even more destructive.

According to some researchers HAARP is utilizing this method and is ready, right this minute, to bring down enemy missiles or spy satellites—or friendly ones!

The website *ColumbiasSacrifice.com* gives the findings of an

204

independent investigation into the Space Shuttle Columbia disaster. That probe was conducted by Jon Hix who received a Bachelor of Science in Mechanical Engineering (B.S.M.E.), with a minor in Aerospace Engineering focusing specifically on the physics of Hypersonic Flight and Spacecraft Reentry, from the California State Polytechnic University, Pomona (CalPoly Pomona—where your author was an "off campus agitator" in the 1960s!). Jon has since worked at McDonnell Douglas Aerospace where he has acquired a great deal of experience in determining failure modes of various mechanical and electronic systems. His research into the Columbia crash uncovered evidence of a massive cover-up within the formal NASA investigation that renders the findings of the official inquiry questionable, at best.

Researcher/engineer Hix concluded that "The Columbia very likely encountered something that destroyed most of its avionics equipment and guidance and flight programs at 13:47:32 during reentry." That is, Columbia was taken out by one sudden electronic event, not a progressive burn-through that disabled one system after another over a period of time. Jon presents evidence that HAARP was used, intentionally or unintentionally, to destroy the Columbia by electromagnetic means!

There are several independent researchers and websites who claim to monitor HAARP and post when they say it is operating. For example, after the terrorist attacks of 9/11 HAARP was reported to be running at full power for the following month. Some bloggers speculated that it was being used to "mind control" America about what had just happened. Others thought that HAARP must have been working in earth penetrating tomography mode, trying to map out the system of caves and tunnels in Afghanistan where Osama bin Laden was believed to be hiding.

Marshall Smith is a former NASA scientist and the President and Director of Research for the Teddy Speaks Foundation, Inc., a non-profit educational corporation. He has degrees in mechanical and electrical engineering, and physics and has long been a licensed radio engineer and senior computer systems analyst. He has worked/consulted for many years on NASA Space Shuttle projects, Titan, Trident and Tomahawk missile systems, and Star Wars laser and particle beam devices. He is familiar first-hand with top secret government, military and industry research facilities. He too maintains an ongoing watch on HAARP. He claims, via his website, Brother Jonathan Gazette (BroJon.com), that HAARP was running in this relativistic particle creating mode for 90 minutes before the space shuttle Columbia attempted reentry and continued for 90 minutes after loss of signal from the orbiter. Smith believes that Columbia was collateral damage, accidentally brought down while destroying a North Korean missile launched at the US!

There is some evidence to support this, in that a nose cone from a North Korean missile may have been found in Alaska after Columbia broke-up on reentry. *The Korea Times*, a major Korean newspaper, ran a story on 4 February 2003 that said a delegation from South Korea's National Assembly had just released a report on the region's showdown with North Korea over nuclear weapons and missiles. The newspaper quoted a former Japanese foreign minister's words in the report that "the last piece of a missile warhead fired by North Korea was found in Alaska." I remember hearing a sound bite news story on the radio at the time. Yet the *LA Times* and the *Anchorage Daily News* have both since ran stories debunking the *Korea Times* report, which can be found on the Web. Marshall Smith insisted, in private correspondence with me, that this debunking of the nose-cone story was a cover-up. Even the *Korea Times* article has been added to since it was posted to the web. Is there a massive cover-up to protect the secret that HAARP is a functional Star Wars defensive shield, or did Marshall just fall victim to a piece of junk reporting?

In an email to me Marshall wrote:

> The public release of Eastlund's "HAARP"

patent in 1991 is itself a part of the disinformation campaign. The use indicated in the patent is for a continuously operating transmitter, using most of the complete oil output from the Alaska pipeline to produce a worldwide radiation shield to prevent missiles from entering the US airspace. But that use is completely illogical since it would also destroy all of the hundreds of satellites now in orbit, along with the ISS [International Space Station] and even all the geosynchronous TV and communication satellites. Obviously not a desirable use of HAARP, so that Eastlund Patent story is a sheer fantasy. Its release to the public was simply military mis-direction and disinformation.

The actual new use, not explained in any public patents is to fire HAARP, not continuously, but in very short 3 second pulses. This is very energy efficient and can produce a spray of radiation which can knock down not one or two but hundreds of missiles simultaneously from China or Korea passing over the Alaska flight path. And if a few of the hundreds of missiles survive the first blast, another blast repeated just 10 seconds later will finish off what is left. No missiles will get through past HAARP coming from Asia.

Also by using short aimed pulses, the HAARP spray blast can hit specific incoming targets directed by AF/NORAD, and also miss friendly satellites and Space Stations. The HAARP output is not like a narrow focused beam or ray gun, instead it is a spreading blast of radiation in space something like the spray from a can of insecticide which can kill and knock down hundreds of mosquitoes at the same time. This, of course, is a problem if the Chinese fire missiles which are lined up with the orbital path of the ISS or other friendly space vehicle.

This was the problem on Feb 1, 2003 when a

Korean missile with a Chinese warhead was fired at central Kansas at the same time Shuttle Columbia was landing on a path just several thousand miles south over Hawaii. NORAD had the dilemma, let the Korean missile pass through, or risk hitting Columbia which was just south of the Korean missile. You know the rest of the story.

Marshall Smith alleges that the North Koreans (under the guidance of the Chinese) fired a missile at the United States after talks with North Korea broke down. He says the Chinese wanted to know how large an area the HAARP defensive shield covered. They got their Korean lapdogs (in a fit of pique) to launch while Columbia was in reentry mode. If HAARP only covered a small area it would destroy the North Korean missile but not Columbia. If HAARP covered a large area the North Korean missile might have gotten through if the HAARP operators had been afraid of taking out Columbia along with the North Korean missile. If there is any truth in this then the HAARP guys either didn't know how big an area the HAARP defense shield actually covers or they decided to loose Columbia rather than risk loosing an American city.

As we saw above one of the things they admitted to be doing with HAARP is "the reduction of charged particle populations in the radiation belts to ensure safe spacecraft systems operations." This could be a really good thing, if the bugs get worked out — and their aren't significant ecological consequences. As you may know the earth is surrounded by the Van Allen Radiation Belts, named after James Van Allen (September 7, 1914 – August 9, 2006) who proved they were there by setting off atom bombs in them in 1958. These belts are areas where the Earth's magnetic field traps radiation, although the source of the radiation is in dispute.

Wikipedia tells us that:

208

The term Van Allen Belts refers specifically to the radiation belts surrounding Earth; however, similar radiation belts have been discovered around other planets.

The big outer radiation belt extends from an altitude of about 10,000–65,000 km and has its greatest intensity between 14,500–19,000 km. ... The inner Van Allen Belt extends from roughly 1.1 to 3.3 Earth radii [The Earth's equatorial radius is the distance from its centre to the equator and equals 6,378.135 km]. ... The gap between the inner and outer Van Allen belts is caused by low-frequency radio waves that eject any particles that would otherwise accumulate there.

Solar cells, integrated circuits, and sensors can be damaged by radiation. In 1962, the Van Allen belts were temporarily amplified by a high-altitude nuclear explosion (the Starfish Prime test) and several satellites ceased operation. Magnetic storms occasionally damage electronic components on spacecraft. Miniaturization and digitization of electronics and logic circuits have made satellites more vulnerable to radiation, as incoming ions may be as large as the circuit's charge. Electronics on satellites must be hardened against radiation to operate reliably. The Hubble Space Telescope, among other satellites, often has its sensors turned off when passing through regions of intense radiation.

Space shuttles, the International Space Station and spy satellites all fly well below the inner Van Allen belt. But to get satellites up into geosynchronous orbit, much less to go to the Moon, Mars or elsewhere in the solar system or beyond, craft have to pass through the Van Allen belts. One of the arguments used by those who say we never went to the Moon is that it was impossible to get there because the radiation would have killed or incapacitated any astronaut who made the trip. Van Allen dismissed these claims. He was also a strong supporter of HAARP, by the way.

If HAARP can reduce the hard radiation in these fields it could be a very good thing for those trying to get into outer space. But there are substantial possible downsides to this. Life on this planet is made possible only because the ionosphere shields us from deadly cosmic rays and solar radiation, much as the ozone layer protects us from ultraviolet light. Some scientists have theorized that the Van Allen belts carry some additional protection against the solar wind (the flow of hard radiation that streams out from the Sun in all directions). Weakening of the belts could harm electronics and organisms on Earth, and, worse may influence the Earth's telluric current (an extremely low frequency electric current that occurs naturally over large underground and underwater areas at or near the surface of the Earth). Dissipating the belts could influence the behavior of Earth's magnetic poles!

Exactly how this reduction of the radiation in the belts is accomplished is not well described in the available literature. It seems to involve inducing the belts to precipitate out (rain radiation to Earth) by injecting either radio waves or excess particles into them. So what happens when this hard radiation falls into the upper atmosphere? Could it create a cascade effect with effects reaching all the way to the ground? What if the attempt to overburden the belt fails and instead of causing it to shed excess radiation it just beefs it up?

DARPA's including "the reduction of charged particle populations in the radiation belts to ensure safe spacecraft systems operations" in the same bullet with "the generation of extremely low frequency/very low frequency radio waves for submarine and other subsurface communication" could mean that doing one also accomplishes the other—that is, could they be implying that one way of using the HAARP antenna array generates radio emissions that do both? Is it possible that while trying to talk to the submarine fleet they might also get the side effect of altering the populations of charged particles in the radiation belts?

210

Jon Hix at *ColumbiasSacrifice.com* rejected Marshall Smith's contention that a North Korean missile was involved in the disaster. He wrote me an email saying:

> All we know is that the warhead stage of such a missile was found in Alaska based on a story from the mainstream Asian press. Unfortunately this story by itself proves nothing. If we had any evidence at all stating when the missile was launched we might have more to go on. If it was launched just a few minutes before the Columbia was destroyed or if we could even pin the launch window down to sometime that morning prior to Columbia's reentry, Marshall's theory becomes more plausible. However, the missile launch could have occurred a month prior to the Columbia accident. We have no way of knowing since neither our government nor the North Koreans are talking. Their silence on the subject is of course not at all surprising.

If HAARP actually were involved in the destruction of Columbia might it have been trying to clear a safe (radiation free) path for the orbiter and something went horribly wrong? Perhaps it was just bad timing and HAARP operators were sending a message to the submarine fleet and had no idea that the orbiter was even there? Describing his conclusion that an EMP or EMP-like event brought down Columbia, Jon wrote:

> The shuttle was designed with triple and quadruple redundant electronics systems as well as other systems such as the hydraulic fluid supply and electric power generation that are divided into three completely separate entities. Based on the extensive maintenance and testing done prior to each mission, the chance of losing all of the systems that perform some function for the shuttle is far too

211

small to calculate accurately. For all of the guidance and control systems to be affected by something that would normally occur during space flight is generally considered impossible, the Columbia had all of the backup systems it needed to make it through the mission. This means that something completely abnormal and unaccounted for affected all of the equipment onboard the orbiter and not just a mechanical or electronic glitch that affected one or two units. Extensive research has led to only two existing possibilities for this scenario. 1.) An Electro Magnetic Pulse (EMP) from a nuclear blast near the Columbia during reentry. This would certainly do it if the missile carrying the warhead were launched at exactly the correct time to intersect the Columbia's flight path. This is not possible, however, according to the accounts written about high altitude nuclear blasts that occurred in the 1960's and 1970's the effects of such an event would have been visible for many miles. There were no such reports of anomalous activity in the sky at that time which would have matched the effects of a high altitude nuclear blast. 2.) The Columbia passed through a region of high speed beta particles intended to mimic the effects of an EMP. All of the academic papers written on the subject state that it is possible to produce such an effect for the purposes of ballistic missile defense. It is known that especially during the Cold War several countries attempted to create such a system to protect them from ICBM attack. The relative success of these systems is unknown and the information can be assumed to be held secret.

Is it possible that the unthinkable happened—that instead of making it safer for Columbia to return to earth, or while engaged in sending a signal to the submarine fleet, or possibly while engaged in some other test involving the shuttle and HAARP, as may have occurred on previous shuttle missions, HAARP accidentally contributed to Columbia's demise?

HAARP AND UFOS

You might be surprised how many times I have been asked about HAARP and UFOs. And oddly enough, there are a couple of possible connections. Of course this is all totally speculative—and you thought the last section was kooky!

Let us start with defining our terms. "UFO" is a horribly misused and abused acronym. It literally means Unidentified Flying Object. Just that—something in the sky that the observer is unfamiliar with. Most "UFO" sightings are at night and often an object cannot be ascertained, making the sighting actually an Unidentified Aerial Light (UAL).

Yet most people using the term UFO have something very specific in mind. They actually have a "provisional" identification, as they think it might be a "nuts and bolts" spacecraft from some other planet (extraterrestrial or ET). They think that because that is the "everybody knows" common explanation given for these things. But where does that explanation come from? Skeptical media reports, sci-fi movies and TV shows, intentional disinformation from military-intelligence sources and often questionable eye-witness accounts!

There are in fact many other explanations for what these UFOs and UALs are, and from whence they originate. These objects could be from this Earth, but are equipment or phenomena unknown to the observer: such as being advanced aircraft; or coming other dimensions (extradimensional); or from other times, like the future (extratemporal); or they may be natural in origin, such as piezoelectric "earth lights."

You may have heard stories that one or more alien craft have crashed and been recovered, by one or more Earth governments, and thus ET technology, after being "back

213

engineered" by our scientists, has been understood and put into use in our daily lives. Another set of stories revolve around one or more Earth governments having made contact, possibly even official diplomatic contact with space aliens and their governments.

One of these famous crash and recovery stories is the event that is said to have happened on 3 July 1947 about 75 miles outside of Roswell, New Mexico. One or possibly two craft, with bodies, were recovered and the remains shipped to Wright-Patterson Air Force Base (AFB) in Ohio for analysis (kept in the infamous Hangar 18). One of the posited explanations for the crash of the craft was that their onboard flight controls were disrupted by the radar from the Roswell AFB (which at the time was home to all of the known nuclear weapons on this planet). Many have speculated that HAARP was built as literal Star Wars, intended not to stop Soviet ICBMs but incoming invaders from the stars.

Others have wondered if HAARP technology comes from the folks who have been back engineering alien technology at Wright-Pat, Area 51, and elsewhere. I have run into numerous reports that a decade before the alleged Roswell crash there was a similar event in Germany's Black Forest in 1936, which may explain World War II Germany's exotic weapons and advanced physics. There are also many rumors that the Nazis made contact with space aliens. Adventures Unlimited Press has released many books and DVDs on advanced Nazi air craft and science, such as *Hitler's Flying Saucers* by Henry Stevens and Joseph Farrell's *Reich Of The Black Sun* and *The SS Brotherhood Of The Bell*, as well as my own *SECRETS OF THE HOLY LANCE: The Spear of Destiny in History & Legend*, co-authored with George Piccard, author of *LIQUID CONSPIRACY: JFK, LSD, the CIA, Area 51, and UFOs*.

Then there are the contact and collusion stories. One of these persistent rumors is the claim that the US inked a treaty with beings from the stars, exchanging our genetic material (from unwilling abductees) for advanced technology (which might have included HAARP). While hundreds, perhaps thousands of people have reported being abducted there is little credible scientific evidence to support these claims. Indeed, science tends to refute these stories.

I think the most telling piece of evidence — curiously both for and against the adduction scenario, depending on how you interpret the data — is the so-called *Persinger Helmet*, a brain research tool designed by Michael A. Persinger Ph.D. He is a registered psychologist in neuropsychology and is employed by the Department of Psychology of the Laurentian University of Ontario, Canada where he is the Coordinator of the Behavioural Neuroscience Program. With his "helmet" he discovered that when specific frequencies are directed into the hippocampus area of the back brain a large percentage of subjects reported UFO abduction experiences, out of body experiences, and a wide range of altered states of consciousness, including "Union with God."

The helmet experience, Dr. Persinger says, "involves a widening of emotional meaning, such that things not typically considered significant would now be considered meaningful" and hallucinations are "perceived as extremely real."

This leads to some very interesting questions. Dr. Persinger assumes that these experiences are entirely hallucinatory, as the research subject remained in the lab under his observation at the same time that the person thought they had been levitated up to or teleported into the alien craft and subjected to numerous indignities. If Persinger is right, what is the origin of these hallucinations? Could it be that they are not hallucinations in the usual sense but are in fact extremely vivid memories? Could they be something in the person's memory, or our collective or racial memory, that the radio frequencies of the helmet are stimulating? Could the subject have been reliving a past event in their life, having been abducted before Persinger experimented on them (and had suppressed the memory or forgotten the event)? Or might the memories be even deeper, from past lives or in our DNA or wherever "instinct" comes from?

But what if the subject really was abducted, not to an alien space craft hovering above a Canadian university, but into another dimension? Could it be that our perceptions or our spirit bodies are shifted just enough by the radio frequencies Persinger is using to make them aware of, and be visible to, entities in another dimensional realm of existence? Could the abducting UFO "aliens" really be extradimensional rather than extraterrestrial? Could Persinger have accidentally discovered a doorway to their "world"?

Dr. Persinger is a civilian scientist and his work has been fully published in the open academic press. Is it possible that military scientist working on "black ops" have replicated Persinger's work, giving them access to that other realm? Is it possible that the stories of one government or another inking deals with space aliens is only half true? Could we be engaged in dialog with entities on this Earth, but from other realms of existence?

And what role might HAARP play in such a scenario? In *HAARP: The Ultimate Weapon of the Conspiracy* I cover at length the science behind the HAARP signal and how it is intended to bathe the world in extremely low radio frequencies at exactly the same frequencies that the human brain works at. HAARP and the US Navy have admitted to doing this as a side effect of communicating with deeply submerged submarines. What if HAARP could put a virtual *Persinger Helmet* on all of our heads all at once? Might we have hallucinations? Or visitations?

In *HAARP* I speculate that HAARP might be used to fake The Rapture in this manner, by getting us to hallucinate Union With God, and thus be directed to accept the false god of the New World Order. But what if these other dimensions are real? Even if done unknowingly and unintentionally, what would happen if we all got irradiated in just the "right" way to shift us over to their dimension? Or even to some higher dimension still? Could HAARP be a mass ascension machine like in the 2000 Brian De Palma movie *Mission To*

Mars?

Or, looking at the negative, what if we are already in the process of "ascending" to a higher spiritual plane—could HAARP's signal act to hold us back? And, what if those controlling HAARP are themselves being controlled by "aliens"? Could Angels or Demons, ETs or extradimensional entities be calling the shots, from behind the scenes, at HAARP? Could they be using it to control us, or each other? Could Angels or friendly ETs use it to boost us up to our next level? Or keep unfriendlies away? Or might Demons or hostile entities use it to do the opposite?

Of course no one in the scientific community is willing to consider such silly "New Age" questions. I considered putting them in *HAARP* then left them out as being just too far "out there." Maybe I should have left them out of this book too…

THE FUTURE OF HAARP

Now that the construction of the FIRI is finally completed it is ready to do whatever the University of Alaska, the Navy and the Air Force want to do with it—which has not been made public. The University of Alaska's website hasn't been updated for years, not since before BAE Systems got the contract to finish it. Working papers that have been distributed among scientists and military planners detailing HAARP plans have not been posted to the Internet, nor as best I can tell, have they been published in the mainstream media.

That is where HAARP is today, a multi-million-dollar "upper atmospheric and solar-terrestrial physics" observatory—that may or may not also be a whole lot more. Remember, it is the High-frequency *Active* Auroral Research Program. There is very little that is passive about HAARP— it is designed to do things, to modify the atmosphere. Like the report "Applications and Research Opportunities Using HAARP" said, HAARP is about making "the transition from pure research to applications in the civilian and military arenas." I think we should be told what those applications will be. Even if HAARP is exactly what it says it is, is that something good?

Initially, my greatest fear was that HAARP would accidentally trigger mental or emotional dysfunction across a large portion of the planet. As I learned more about HAARP an entire Pandora's Box of potential dangers came to view. Clearly, HAARP is a ground-based "Star Wars" weapons system. The Soviets objected to SDI on the grounds that it would give the US First Strike capability. That has not changed, although the dynamics of the Cold War has. HAARP still has the potential to destabilize international relations, only now with the Chinese, rather than the FSU (Former Soviet Union).

From an environmental point of view there are no end to the potential horrors of HAARP and similar technologies. Oddly enough, my greatest fear now is that HAARP might be exactly what they say it is — an experimental base from which clueless civilian scientists are manipulated into performing questionable experiments on vast planetary systems with unknown potential outcomes — even if they don't say it in quite those words.

And why are these experiments being conducted? HAARP's primary avowed purpose, stated most simply, is to find out if the atmosphere can be made into some kind of weapon; or if portions of it, such as the ionosphere, can be "tweaked" so as to advance military objectives. Common Sense tells us that "weaponizing the atmosphere" could be a serious mistake — even if it works. If we achieve this technology how long will it be before our enemies gain it too? What happens when we and they go head to head in a war using geophysical weapons? Nukes are nice in comparison to the hell that could be unleashed.

Americans, to their detriment, demand simple answers to complex questions. HAARP has been used as the "Ah ha! That's the reason!" for just about everything that cannot be otherwise easily explained: from twisters out of season to the Asian Christmas tsunami; from multi-state power outages to voices in one's head. Since HAARP burst into public consciousness the public has been too willing to buy HAARP as the explanation for all manner of things that went bump in the night. Too many have latched onto HAARP as The Answer. It's not. HAARP is real and HAARP is dangerous, but HAARP is probably not what most people imagine it to be. There are literally millions of other research projects going on around the world, of which most people know nothing. In the tens of thousands of projects being conducted by DARPA alone right this minute are many that make HAARP look like a child's toy, but you don't hear about them. Connecting the dots is good—but one has to possess the correct dots first.

HAARP is a symptom of a bigger disease, the victory at any cost, ours to command and control philosophy that drives the military-industrial-academic complex to create these ever more terrifying monsters of technology, destroying our soul while trying to defend our stuff. It is that diseased philosophy that we need to drag out into the open and expose to the light of reason.

What other outrageous and dangerous experiments are going on that we don't yet know about? What horrible new threats to our environment, our liberties and our way of life are going on hidden away on secret military reservations, like Area 51, or just down the block at your local university or across town in an industrial facility? What might be going on under our very feet or *in the skies over our heads?*

[handwritten margin note: GOOD QUESTION ?]

CHAPTER SIX

CHEMTRAILS?

Have you looked at the sky recently? Have you noticed anything different? What about the contrails left by aircraft, do they seem odd to you? Some people say that the contrails have changed and call them *chemtrails*. They say that there is some kind of conspiracy being engaged in to hide the fact that these new and different chemtrails are the visible phase of a high-altitude spraying program. There are many rumors as to what is being sprayed and why.

One of the claims is that the sky is being laced with heavy metal particles to counteract the effects of global warming. Edward Teller, the father of the hydrogen bomb, proposed doing this so there may be something to it, which we will examine in detail in the next chapter.

Some of the people who have posted on this subject to Internet websites and discussion groups say that these sprays are not of chemical but of biological substances. Some optimistically believe that these sprays are of drugs meant to secretly immunize us against biological agents that might be released by terrorists. Whereas others claim to know that these high altitude sprays are of bio-hazardous materials intentionally released to make us all sick, killing off the old and weak, culling the human herd in a plan to reduce the population to levels the elite believe "sustainable."

Other postings state that when these substances fall to Earth they alter the alkalinity or electro-conductivity of the soil so that only genetically modified foods will grow, passing profits to the multinational agribusinesses like Monsanto and Archer Daniels Midland (ADM) by wiping out "heritage" crops. And still wilder claims are made that these sprays are part of a covert mind control system or that they are either to protect us from invading space aliens or make this world more habitable for them.

One website (lightwatcher.com) rather neatly summed up the theories as to why this alleged spraying is taking place, with the author ranking them in order from what he considered the most likely to the least likely reasons:

1. To mitigate global warming and the rapid collapse of Earth's ecosystems, to patch the ozone holes, and to shield life from the increased infrared, ultraviolet and cosmic rays from space.

2. For global weather modification—utilizes cloud making and seeding to control precipitation.

3. To facilitate electromagnetic mind control technologies that subjugate and control human populations physically, mentally and spiritually.

4. For Defense applications: concealment from aerial and satellite observation, to facilitate a new military communications system, to defend against incoming ICBMs (when combined with ELF and EMP waves).

5. To facilitate an advanced military 3D imaging system.

6. To work with a variety of the uses listed above, as 'Operation Cloverleaf' is quite versatile.

7. As a delivery system for mass immunizations without the knowledge of the population.

8. In conjunction with HAARP array to sterilize areas contaminated with biotoxins and toxic molds.

9. As a carrier for mind control transmissions from U. S. governmental sources.

10. To shield the U. S. from ELF mind control transmissions from Russia and China.

11. As a delivery system to facilitate global population reduction to sustainable levels.

12. To geoengineer Earth's atmosphere to

accommodate extraterrestrial invaders.

Chemtrail investigator Diane Harvey summed it up like this:

> It is beginning to look as if the actual purpose of all that we have been investigating is nothing less than the actual physical transformation of the earth's atmosphere in order to provide a platform for the latest chemical/electromagnetic technologies of warfare, communication, weather control, and control of populations through non-lethal chemical electromagnetic means. And what this portends for the future in terms of any meaningful retention of human freedom, and even the very life of the planet itself, is entirely unknown. Therefore it is not beyond reason to suggest that, unless this project is forced into the light of public scrutiny by means of a relatively few dedicated citizens, human freedom itself and perhaps even all life on this planet may be at risk.

Can any of these crazy sounding claims be true? Unfortunately, yes. Publicly available documents show that both civilian and academic organizations and agencies of several governments from around the world have been engaging in injecting substances into the atmosphere for decades, such as silver iodide and barium as we have already seen. These chemical releases have been from balloons, aircraft and rockets, as well as from stationary platforms on the tops of mountains.

Civilian operations have mainly been for weather modification, such as cloud seeding to induce rain and to prevent hail from forming and destroying crops. Literally hundreds of academic and scientific institutions have taken part in releasing substances into the atmosphere in efforts to better understand the characteristics and processes thereof. The military has many release programs engaged in a wide variety of communications and surveillance activities. These include releasing radar jamming "chaff" during training exercises and various experiments and operations in developing and using advanced communications and imaging technologies.

Although some, but by no means all, of these releases are kept under a cloak of secrecy for "national security" reasons, most people would accept that these are legitimate, although possibly dangerous activities expected of research scientists and military operations. But what if it's more? Could the white plumes trailing behind jets really be symptoms of some dark conspiracy?

A *contrail* is a trail of condensation (conversion of water from a gas to a liquid) that is left behind by a passing airplane, most commonly a commercial jet (military craft try to avoid leaving them to remain undetected and piston-driven aircraft seldom get high enough to reach air that is cold enough to cause contrails to form).

A *chemtrail* is an alleged plume of intentionally released chemicals or other material, either sprayed from nozzles, or as particulates mixed with the plane's fuel that mimics a contrail. Many people around the world are concerned about this purported spraying program. Others are convinced that chemtrails are nothing more than urban legend and hysteria, with perhaps a touch of disinformation.

Frankly, I find the arrogance of those who insist that there is nothing to the chemtrail stories amazing. The only way that one could know with absolute certainty that nothing was going on would be to know literally everything that was going on with every aircraft on the planet, and there is no one that omniscient. Those who insist that it is all nonsense can only be speaking from their ignorance of any specific project and their arrogance in assuming that they are so important that if anything were going on those doing it would surely tell them all about it! Well, as the UFO folks put it, absence of proof is not proof of absence!

There is a growing body of scientific evidence that contrails are hardly benign, as will be covered in this chapter—but are they part of some global plot? There is circumstantial and documentary evidence that there are, or may be several "spraying" programs both civilian and military in nature—but are they criminal conspiracies? Those who have amassed this evidence believe that there are indeed one or more conspiracies involved. Critics dismiss it all as insanity and "conspiracy theory."

Calling something a "conspiracy theory" is a propaganda technique. Conspiracy theories are only theories until they are proven. In the United States conspiracies are proven in the courts week after week. People constantly conspire to traffic in illicit drugs, child pornography, or to commit other crimes. Why should we find it so difficult to accept that individuals within big corporations (like Enron) or agencies of our government (like the CIA) might conspire to conduct shady business?

It is a proven fact that civilian contractors over-bill and otherwise steal millions from the government. As I write this, the jury has just found former Illinois Governor George Ryan Sr. guilty on all charges of corruption and conspiracy, proof that officials at the highest levels of government do, at least occasionally, conspire to commit crimes. Check our prisons and you will find that the majority of those incarcerated were convicted of "conspiracy to..." in addition to other charges. Yes, conspiracy is a growth industry in America today, yet lazy journalists and apologists for the status quo dismiss anything not readily provable as "conspiracy theory." Or, as Gore Vidal put it: "Apparently 'conspiracy stuff' is now shorthand for unspeakable truth."

It is also a known fact that governments engage in covert activities, some of which may be completely legitimate, but not something they would want foreign powers to know about. On the other hand, some of these covert ops are highly questionable, such that citizens and other governments would object if they were revealed. Chemtrails could be examples of both.

Delving into the possibility of clandestine operations and/or "conspiracies" is the time-honored role of the investigative journalist. Where would Woodward and Bernstein be if they had stopped looking into Watergate just because someone said it was all a "conspiracy theory"?

So, what can we prove about these chemtrail theories? Let us start with understanding contrails in general and the environmental hazard they pose.

WHAT IS A CONTRAIL AND HOW DOES IT FORM?

Contrails mostly form when humid air from an aircraft's engine exhaust mixes with air of low vapor pressure and low temperature. Vapor pressure is a term for the amount of pressure that is exerted by the water vapor itself (as opposed to atmospheric, or barometric, pressure which is due to the weight of the atmosphere). If the air is sufficiently moist contrails can also form because of the pressure differences between air streaming over the wings meeting air coming under them. If condensation does occur, then a contrail becomes visible, either immediately behind each engine or the plane as a whole (usually behind the tail) or sometimes just from the wing tips.

When hydrogen in hydrocarbon-based jet fuel is burned it combines with oxygen in the air yielding water (H_2O) that is emitted from the engine. This water quickly freezes into ice crystals since air temperatures at high altitudes are very cold (generally colder than minus 40°F). Because of these low temperatures only a small amount of liquid is necessary for condensation to occur. If temperatures are above minus 40°F or the relative humidity of the air is less than 73%, it is unlikely that a contrail will form. The ice crystals forming the contrail behave like a naturally made cloud. If enough moisture is already present in the air, the contrail can spread by growth of the ice crystals.

226

ARE PERSISENT CONTRAILS ACTUALLY CHEMTRALS?

I really hate to go up against this popular misconception, but the mere persistence of a contrail is no proof of its being a chemtrail. Before you throw up your hands and say I'm part of the conspiracy, please read a little further. There are chemtrails, as I will show, but the persistence of a contrail is not a reliable indicator of its contents—a persistent contrail may or may not be a chemtrail.

The artificial formation of clouds via contrails is very similar to the process that occurs when you breathe on a cold winter day and you can see your own breath in the form of a "cloud." You may have noticed that on some days this cloud you produce lasts longer than on other days. The length of time that a contrail lasts is directly proportional to the amount of humidity that is already in the atmosphere. Drier air leads to shorter-lived contrails, whereas high humidity will lead to longer-lived contrails—with enough moisture all jet engines will produce contrails with or without a "secret ingredient." If the atmosphere is too dry, no contrails will form—so even if a plane actually were spraying, you wouldn't see it!

The skies above the Southwest are typically too dry, and the skies above the Deep South are generally too hot for extended contrail coverage. These factors plus the varying density of air traffic over different parts of the country combine to make the skies in the Midwest and Northeast and, to a lesser extent, the Pacific Northwest particularly laden with contrails.

Weather watchers have known for some time to look at the persistence of contrails for clues in weather forecasting. On days where the contrails quickly disappear or never form, one can expect continuing good weather as the air aloft is very dry, while on days where contrails persist a change in the weather can be expected. Natural cirrus clouds are the first harbinger of an approaching storm, appearing 6

227

to 24 hours before the bad weather arrives. Artificial cirrus from contrails indicates moisture-laden air at high altitude. Contrail "cloudbanks" can form a day or two ahead of the natural cirrus giving farmers and backpackers an early heads up on what might be coming. The "persistence" of a contrail, then, is not a good indication of its contents.

Occasionally a jet plane, especially if ascending or descending, will pass through a much drier or more moist layer of atmosphere which may result in a broken pattern to the contrail, with it appearing in segments rather than in one continuous plume. This can give the appearance of "spray nozzles" suddenly turned on or off. Of course, it is possible that sprayers actually are being turned on or off and the atmospheric conditions give "plausible deniability."

ARE PERSISTENT CONTRAILS A RECENT PHENOMENON?

As mentioned earlier, as early as 1953 Herbert Appleman of the U.S. Air Weather Service noted that contrails could lead to the formation of clouds that could spread out and persist for extended periods. Good old-fashioned contrails have been known to persist for many hours, in some cases even days, if enough ambient moisture is available to inhibit their dissipation.

Despite dire Internet postings persistent contrails are not a recent phenomenon. Ever since airplanes have been able to reach air cold enough to permit them to form (minus 40°F and below), possibly as early as the Spanish Civil War in the 1930s, contrails have been forming and persisting. Many reports exist from World War II of situations where the accumulations of contrails were so extensive that pilots were unable to keep visual contact with neighboring or enemy planes during combat. Persistent contrails were a significant threat to the stealth of bomber squadrons. Occasionally alternate return routes had to be taken to avoid contrail cloudiness left behind on the outbound flight (wherein enemy fighters might be waiting in ambush).

I have been watching old movies lately trying to find when the giant billowing persistent type of contrail/chemtrail we have today first turns up on film. The earliest example of what so many people call *chemtrails* that I have been able to find is in the opening scene of the 1985 film *Murphy's Romance* with Sally Field and James Garner. I have found many films with these so-called chemtrails from the early 1990s, so whatever they are they did not start a year or two before the turn of the 21st century as so many ill-informed people claim.

Persistent contrails, then, are not automatically chemtrails, unless they all are, which is somewhat unlikely. Conversely, just because you don't see any contrails doesn't mean that the air above you isn't being tainted with strange chemicals!

I first heard the term *chemtrail* in 1995. At the time I was working with Jim Keith in researching a number of his books. We pulled together a circle of researchers, media watchers, political activists and paranoids to assist us. Oddly, that circle, the Saturday Night "Eaters" group, continues to meet to this day, currently holding forth in the banquet room at the Denny's in Sparks, Nevada, every Saturday night from 6 PM till around 10 PM – and yes, you are invited! I first heard the term from one of our diners. It soon became all the buzz.

I didn't put much stock in it till the following summer. In 1996 I got the go-ahead on my first non-fiction book, *HAARP: The Ultimate Weapon of the Conspiracy*. I spent much of that summer lying in a lawn chair in my garden in Reno, Nevada, thinking about what I was going to write next and watching the sky. Yup, I saw 'em. Big fat contrails filling the sky, sticking up there for hours, blending into each other, turning the sky a milky opalescent white by late afternoon. They were like nothing I could remember seeing before. These were most certainly not the pretty short-lived contrails of my childhood in dry, sunny Southern California. At the time I thought that either the engines had changed, or the fuel had changed, or the atmosphere itself had changed. Actually, it was all three!

229

CONTRAILS AND GLOBAL WARMING

Global warming has been of great and increasing concern to environmental scientists and international bodies like the United Nations for a few decades now. Some scientists, but not by any means all, believe that the so-called greenhouse effect is responsible for global warming. The argument is that ever-increasing amounts of heat-trapping gases have been generated since the beginning of the Industrial Revolution. These gases, such as carbon dioxide (CO_2), chlorofluorocarbons (CFCs), and methane accumulate in the atmosphere and allow sunlight to stream in freely but block heat from escaping, acting like the panes of glass in a greenhouse.

By examining ice cores, tree rings, and fossils, scientists have estimated that the northern hemisphere is warmer now than at any time in the past 1,200 years. In November 2005 *Science* magazine reported on research into Antarctic ice cores that showed carbon dioxide levels are now 27 percent higher than at any point in the last 650,000 years.

The scientific community is generally in agreement that for the past 400,000 years the concentration of CO_2 in the atmosphere has fluctuated between about 180 and 280 ppm (parts per million, the number of individual CO_2 molecules per one million molecules of air). Ice cores from around the world seem to show that beginning in the late 1800s atmospheric CO_2 increased from about 280 ppm to the current level of almost 370 ppm. Many scientists and other (largely self-appointed) spokespersons for humanity blame this sharp rise in atmospheric CO_2 on humans burning fossil fuels and the flatulence and fecal effluvia of certain domesticated animals (atmospheric methane is primarily attributed to cows).

The Intergovernmental Panel on Climate Change (IPCC) is headquartered in Geneva, Switzerland. It is an international group of scientists that was convened more than a decade ago under United Nations auspices to report regularly on the problem of global warming. In 2005 they reported that the average level of carbon dioxide in the Earth's atmosphere was 367 parts per million (ppm), up from 315 ppm in 1957. The IPCC believes that if we fail to control emissions the atmospheric CO_2 level will continue to rise, reaching between 450 and 640 ppm by the year 2050 and possibly twice that by the end of the century. As a result the IPCC projects that by the year 2100 the average rise in global temperature will be between 1.4 and 5.8 degrees Celsius relative to the 1990 level. This projection is based on the IPCC's conviction that there is a link between atmospheric CO_2 (the most common of the so-called greenhouse gases) and global warming. These scientists say that as levels of CO_2 rise the Earth gets hotter because of the greenhouse effect — but not everyone agrees with that assessment.

While the jury is still out on how this is occurring there is little doubt that disaster would be the norm should global temperatures increase into the upper range of predictions. With global temperature increasing so too would deaths from heat waves, failing crops, infectious diseases, and so on. People already facing food shortages could be particularly at risk as for each one-degree-Celsius increase in temperature above optimal levels, wheat, rice, and corn yields are projected to fall by at least 10 percent. Even at the low end of temperature increase guesses climate change models predict more frequent and more severe storms, floods, heat waves, and droughts.

Global warming has been identified by the National Oceanic & Atmospheric Administration (NOAA), an agency of the United States Department of Commerce, as responsible for abnormally high temperatures in the North Atlantic Ocean in 2005 that resulted in disasters both above and below the ocean's surface. With the article "Caribbean coral suffers record die-off," Cable Network News (CNN) reported on Friday, 31 March 2006 that researchers were appalled to discover that within the previous three or four months several species of Caribbean corals had suffered from an unprecedented rate of mortality. "Early conservative estimates from Puerto Rico and the U.S. Virgin Islands," they said, "find that about one-third of the coral in official monitoring sites has recently died."

CNN cited sea surface temperature figures from NOAA that showed sustained heating in the Caribbean during the summer and fall of 2005 to be the worst in the 21 years of satellite monitoring. The news story revealed that the situation for coral was even worse in some areas of the Indian and Pacific oceans where mortality rates had been in the 90 percent range.

In his article "Coral Reefs And Marine Life May Be Wiped Out By Global Warming," Rod Minchin reported in *The Scotsman* newspaper's Sunday, 21 May 2006 edition that a Newcastle University-led team had just published its findings in the *Proceedings of the National Academy of Sciences*. The team was comprised of researchers from Newcastle University; the Australian Institute of Marine Science in Townsville; the Centre for Environment, Fisheries and Aquaculture Science, Lowestoft; the Seychelles Centre for Marine Research; and the Seychelles Fishing Authority. In this report Minchin said that:

> The international team of researchers surveyed 21 sites and more than 50,000 square metres of coral reefs in the inner islands of the Seychelles in 1994 and 2005.
>
> Their report is the first to show the damage of global warming on the inner Seychelles coral reef in which rising sea temperatures have killed off more than 90 per cent of the coral.
>
> The research showed that, while a warming-up of the Indian Ocean in 1998 was devastating in the short term, the main long-term impacts are down to the damaged reefs being largely unable to reseed and recover.
>
> Many simply collapsed into rubble that became covered by unsightly algae.

The collapse of the reefs removed food and shelter from predators for a large and diverse amount of marine life—in 2005 average coral cover in the area surveyed was just 7.5 per cent.

The survey showed that four fish species are possibly already locally extinct, and six species are at critically low levels.

The survey also revealed that species diversity of the fish community had decreased by 50 per cent in the heavily impacted sites.

Tom Goreau of the Global Coral Reef Alliance has calling what's happening worldwide "an underwater holocaust."

The rise in sea surface temperature also disastrously contributed to 2005's record-breaking Atlantic hurricane season, with 27 named storms and 15 hurricanes. As you probably know, in August 2005 Hurricane Katrina devastated the U.S. Gulf Coast, leading to more than 1,100 deaths and displacing approximately 1 million people.

Established in 1998, The Pew Center on Global Climate Change declares itself as one of the leading purveyors of information leading to action on global warming. Although they claim to be non-partisan The Pew Center is firmly in the "greenhouse gases cause global warming" mind-set and is a key supporter of "sustainable development." In its first seven years in existence the Pew Center issued 92 reports on such climate change topics as economic and environmental impacts and domestic and international policy recommendations. The Pew Center distributes its reports to more than 4,000 opinion leaders throughout the world with thousands more downloaded monthly from the Center's website. The research is regularly featured in major news stories from the Associated Press, *Nature Magazine*, *The New York Times*, and other media.

The Pew Charitable Trusts as well as corporate and private donations support the Pew Center on Global Climate Change. An indication of how far up the globalist food chain its support base rests can be seen in one of its offshoots, The Pew Center's Business Environmental Leadership Council (BELC), which is composed mainly of Fortune 500 companies. BELC members, they proudly proclaim, generate over $1.5 trillion in revenue and employ more than 2.5 million people and include General Electric, Alcoa, and DuPont.

The Earth Policy Institute, founded by Lester R. Brown, who was called "the guru of the global environmental movement" by *The Telegraph of Calcutta*, is a small but vocal organization pushing the "greenhouse gases cause global warming" agenda. Its purpose is given as: "to provide a vision of what an environmentally sustainable economy will look like, a roadmap of how to get from here to there, and an ongoing assessment of this effort, of where progress is being made and where it is not."

Joseph Florence, one of the few full time staff at the Earth Policy Institute, regurgitated one of the Pew Center reports in his article "*2005 Hottest Year on Record*." The Pew Center on Global Climate Change had analyzed the results of 40 previous studies and found a clear link between increased temperatures and numerous changes in natural systems across the United States. Florence reported that Pew's scientists found that:

> Warmer winters, increased precipitation, and earlier springs are causing certain plant species to bloom several weeks earlier, which is disrupting insect food supplies and plant pollination cycles. Temperature changes have led to shifts in many species' habitats as populations move north and to higher elevations in search of cooler temperatures. Scientists estimate that about half of all wild species in the United States have already been affected by climate change.
>
> Warming in the Arctic — the area around the North Pole, including parts of Russia, Alaska, Canada, Greenland, and Scandinavia — has occurred at nearly twice the global average rate. Indeed, a snapshot of 2005 shows that the greatest warming last year occurred

in the Arctic Circle. Warming there is enhanced by a positive feedback mechanism. Snow and ice reflect some 80 percent of solar radiation. When they melt, more heat is absorbed by the underlying surface, which in turn melts more snow and ice. From 2002-2005, summer Arctic sea ice has covered 20 percent less area than its 1978-2000 summer average. The Arctic could be ice-free in summers by the end of this century, threatening the fate of the polar bear as melting ice shrinks its habitat and compromises access to food.

In addition, in Western Siberia, an area of permafrost spanning a million square kilometers — the size of France and Germany combined — has recently begun to melt for the first time since it was formed over 11,000 years ago at the end of the last ice age. This permafrost covers the world's largest frozen peat bog. Scientists warn that if warming trends continue it will release billions of tons of stored carbon into the atmosphere, accelerating global warming.

Many scientists believe that global warming is real and serious, although some are concerned about uncertainties in the science. One such was Roger Revelle, the late professor of ocean sciences at Scripps Institution of Oceanography. Although he was a supporter of the green house gases theory of global warming he coauthored, with S. Fred Singer (of whom we will hear more in the next chapter) and Chauncy Starr, a paper recommending that action concerning global warming be delayed insofar as current knowledge was totally inadequate. Another active advocate of global warming, Michael McElroy, head of the Department of Earth and Planetary Sciences at Harvard, has written a paper acknowledging that existing models cannot be used to forecast climate.

A sizable number of scientists accept that the Earth at least seems to be getting warmer, but seriously question if any human activities are the cause, much less green house gases. Although the mainstream press makes it seem like there is a consensus among scientists about the cause of global warming and its danger, that may not be the case. One scientist who has written extensively against the global warming hypothesis is Richard S. Lindzen, the Alfred P. Sloan Professor of Meteorology at the Massachusetts Institute of Technology. He began his article "Global Warming: The Origin and Nature of the Alleged Scientific Consensus," with:

> Most of the literate world today regards "global warming" as both real and dangerous. Indeed, the diplomatic activity concerning warming might lead one to believe that it is the major crisis confronting mankind. The June 1992 Earth Summit in Rio de Janeiro, Brazil, focused on international agreements to deal with that threat, and the heads of state from dozens of countries attended. I must state at the outset, that, as a scientist, I can find no substantive basis for the warming scenarios being popularly described. Moreover, according to many studies I have read by economists, agronomists, and hydrologists, there would be little difficulty adapting to such warming if it were to occur. Such was also the conclusion of the recent National Research Council's report on adapting to global change. Many aspects of the catastrophic scenario have already been largely discounted by the scientific community.

Unfortunately for us this is no mere academic debate. The politics and economics of the 21st century will turn on how this debate plays out — and our survival as a species may hang in the balance.

Ken Caldeira is an atmospheric scientist at the Lawrence Livermore National Laboratory (LLNL) and one of those convinced that global warming is real and must be dealt with immediately. "The question now," he says, "is what can we actually do about it?"

Chemtrails were Lawrence Livermore's answer to that question! Incredibly, since the 1970s several scientists and scientific bodies have seriously proposed adding tiny particles of metallic oxides and other things to jet fuel to intentionally form artificial clouds (real chemtrails) for a variety of reasons, which I will explore in depth in the next chapter. This you will soon see is not a *theory* but a *strategy*. The only question is has anybody actually done it.

The warmer the air is the more moisture it can hold. If global warming is real then the atmosphere should be warming up, if only by a degree or two, which means it should hold more moisture. This would have far-reaching effects. More moisture added to storms would create both more storms and more powerful ones too. This also means increasing drought as the moisture that should fall as rain is being held in suspension aloft, and may be transported away by wind currents and deposited as rain somewhere that has already had too much — i.e., flooding and drought are seen as different sides of the same global warming coin.

Could the wicked weather we have seen for the last several decades, and the last few hurricane seasons in particular, be evidence of global warming? The problem is, so far atmospheric moisture levels have remained constant, or have even decreased over portions of the globe, as I will show shortly. The only thing that seems to have changed is more clouds — more of these strange "chemtrail" clouds — over high air traffic regions of North America and Europe. From whence did those clouds come?

THE HIGH BYPASS TURBOFAN

The first generation of jet engines were used from the 1950s to the 1980s. The current generation began appearing in the early '80s and the conversion to them was completed by 1990 or so. In email correspondence with me the previously mentioned Marshall Smith from the Brother Jonathan Gazette (BroJon.com) described it to me like this:

All of the first generation jet engines had the appearance of a short cigar with the inlet nozzle and the outlet exhaust nozzles about the same size, about 3 feet in diameter. The new next generation jets look like the same old jet but somebody has strapped on the front of it a huge fan about 20 feet in diameter right on the old engine. All the present day jet engines have the front nozzle opening about 6 times larger than the exhaust in the back.

This new addition is called a "High Bypass Turbofan." It is like a turbocharger add-on for a car. It pumps in extra air into the jet engine. But most of the air flows around and past the engine. This maintains a constant sea-level air pressure at the front of the engine even during high thrust takeoff and also in the extremely thin air six miles up at normal cruise level.

The new generation engines never "starve" for air and thus never dump half-unburned smoky fuel out the back. This increases the efficiency of the engine, produces more power and eliminates all the sooty oily black exhaust especially during take-offs and at high altitude cruises at 36,000-foot flight level.

Most of the people who believe in "chemtrails" are all over 40 years old, since they are the only ones who can remember back in the day when all jets took off in clouds of black smoke at the airport, and left long brown exhaust streaks across the sky at high altitude. That doesn't happen anymore, since the early 1980's. Young people today have no clue that "chemtrails" seemed to appear about the same time as the smoky dirty jet planes disappeared.

The primary purpose of the invention of the next generation jet is to increase the efficiency of the powerful jet by a rather large 25 percent. This means it takes 25 percent less fuel to fly an airliner from point A to B, thus a huge 25 percent increase in profit for the airlines, and at the same time lowering the cost of air travel. Plus the same plane can now fly 25 percent farther on the same tankful of fuel. That is a large change in jet engine performance just by strapping on a simple fan on the front of the engine.

This is a substantial improvement and boon to the airline industry. And all just by slapping a big fan on the front of the old jet engines. When this simple technique was discovered in the late 1970's, all existing jet planes were quickly retrofit with the strap on modification to the front of the old jet engines.

One of the first new production-line planes with the modified engines was the Boeing 767 in 1981. Soon even all the older planes with the old-style engines were modified with the strap on "High Bypass Turbofan" added on the front. This includes all the old military tankers, such as the KC-135's which are military version Boeing 707's. Even all old Boeing 747's were quickly modified in the early 1980's starting with the Presidential Plane, Air Force 1.

By 1990, all old jets had finally been upgraded to the new engine configuration. Why would anybody continue to fly a jet which simply dumps 25 percent of the unburned fuel out the back in the form of oily sooty exhaust? The fuel savings alone paid for the cost of the modified upgrade in just the first few flights of the plane. The modified next generation jet engines are now universal and a definite boon to anybody who flies jets.

But the new next generation jet engines have one curious "side effect." They produce contrails which may persist high in the sky for more than 24 to 36 hours. And there is an obvious reason for this. Most people who claim to see "chemtrails" usually claim they know what contrails are. They only last for about 10 or 20 minutes and then they disappear. But these new "chemtrails" persist all day long, and with thousands of flights in the US air lanes each day, these new persistent contrails produce numerous criss-crossing, tic-tac-toe patterns all across the sky. So why do the new contrails now last all day long?

In the old-style sooty exhaust jets, and even old piston engine planes of WW II, the high altitude hot exhaust produced freezing ice crystals which usually formed around tiny droplets of the oily exhaust soot particles. This was like a small dirty snow crystal with an oily spot at the center. The black oily spot quickly absorbed the heat from the sun and caused the little ice crystals to melt within several minutes. When the ice crystals melt, they disappear. The old contrails usually only lasted for several minutes.

But the new ultra-clean exhaust of the next generation jet has no black oil spots in the center of the contrail ice crystals. The heat and light from the sun pass right through the clear ice crystals of the next generation contrails as if they were transparent. They do not absorb any solar heat and thus do not melt. They may remain frozen and persist looking like long thin high cirrus clouds for 24 to 36 hours before they simply disperse but do not melt and disappear.

I have known about this since 1981. I was then an aeronautical engineer at NASA-Ames when I first discovered this odd effect of the new-style high bypass jet engines. So why doesn't anybody else know about this? Nobody asked me.

As Marshall mentioned above, in addition to contrails forming from water emitted from the engine and freezing, tiny particles within the exhaust also provide surfaces for the ice crystals to form on. These particles, broadly called aerosols, are droplets of sulfuric acid, specks of soot and other substances that serve as seeds called condensation nuclei.

Dr. Bruce Anderson was an atmospheric scientist at NASA's Langley Research Center before recently being named Chief Operating Officer of The National Space Science and Technology Center in Huntsville. He was part of a team of scientists that actually went aloft in April and May of 1996 to follow in the turbulent wakes of jets leaving contrails to take samples of the exhaust constituents. He developed a theory for the role of contrails in causing atmospheric moisture to condense into clouds. "We do know that the sulfur in the fuel from aircraft generate aerosol particles, and those in turn can influence the formation of clouds."

Dr. Anderson was part of a NASA mission with the self-congratulatory name of SUCCESS, for Subsonic Aircraft: Contrail and Cloud Effects Special Study. In flights over the Central United States, the Rocky Mountains, and the Pacific Ocean, the SUCCESS team measured emissions of sulfur and soot, with the aim of understanding how these affect high-altitude clouds.

The SUCCESS tests were reported on in July of 1996 in "Ten Thousand Cloud Makers: Is Airplane Exhaust Altering Earth's Climate?" by Richard Monastersky. He wrote:

> By studying what happens to engine exhaust immediately after it leaves the plane, SUCCESS aims to reveal how sulfuric acid and soot alter clouds. Although participants in the project are only now beginning to sift through the data, the sulfuric acid measurements have already shown some surprises. Previous engine tests conducted on the ground had suggested that most of the sulfur emitted by jets comes out as gaseous sulfur dioxide, with less than 1 percent in the form of sulfuric acid. But SUCCESS observations made at cruising altitude indicate that at least 10 percent of the sulfur in the exhaust appears as sulfuric acid droplets, making jet pollution an efficient

producer of clouds.

Project SUCCESS was launched to address the question of whether aircraft emissions are increasing the number of clouds and/or are altering atmospheric chemistry. Project scientist Randall R. Friedl of NASA headquarters in Washington, D.C., recognized that either of these could affect the weather down on the ground, adding: "There are 10,000 large-size commercial aircraft in operation today. It's expected that this number will double by the year 2020. It's a natural question to ask whether these are having an environmental impact."

Ironically, technical advances in engine efficiency have resulted in jet engines that burn fuel more completely, thus combining more hydrogen with oxygen and yielding more water for contrail formation. Better engines also have resulted in cooler exhaust temperatures, making it easier for the contrails to form. The prospect of a hydrogen-fueled jet that would be environmentally clean (zero emissions) but would leave a massive contrail of water vapor would surely raise aesthetic questions — if not charges of conspiracy!

GLOBAL DIMMING

Scientists have also discovered that each year less sunlight reaches the surface of the Earth. *ABC News* reported in "Are Skies Dimming Over Earth?" on February 9, 2006, that:

Scientists have long argued that human activity may be warming the Earth through a process known as the greenhouse effect. Now studies show we may also be having a different kind of impact: global dimming. Researchers have found the amount of sunlight hitting the ground over China dropped by 3.7 watts per square yard over the last 50 years.

Since 1960 10% less sunlight has reached Earth's surface. Levels of solar radiation reaching parts of the former coal-belching Soviet Union are down almost 20%.

The PBS television show *Nova* broadcast a documentary on this phenomenon on 18 April 2006 under the title of "Dimming the Sun." That show reported on:

> ...the discovery that the sunlight reaching Earth has been growing dimmer, which may seem surprising given all the international concern over global warming. At first glance, less sunlight might hardly seem to matter when our planet is stewing in greenhouse gases. But the discovery of global dimming has led several scientists to revise their models of the climate and how fast it's changing. According to one recent and highly controversial model, the worst-case warming scenario could be worse than anyone has predicted.

From the website for the "Dimming the Sun" broadcast we read:

> For millennia humans no doubt have noticed that smoke and ash from volcanic eruptions can block sunlight for many days. But Benjamin Franklin went a step further in 1783, proposing that a massive volcanic eruption of the Laki fissure in Iceland caused months of unusually cold weather in Europe. By the early 1900s, scientists had begun trying to quantify how volcanic eruptions affect climate, but measurements and climate models were too crude to conclusively link the two.
> Early computer models of the global climate attempted to factor in aerosols. It was a daunting task: A wide spectrum of aerosols exists in the atmosphere — small sulfate particles, salt crystals from the oceans, soot, and many others. How these particles, at various

heights, cause absorption or reflection of the sun's radiation was poorly understood. Yet different groups of modelers came to the same tentative conclusion. Human-made aerosols, they found, were contributing to cloud formation, increasing the planet's reflectivity, and causing a modest cooling. Some scientists even suggested that air pollution, if unrestrained, might trigger a new ice age. Yet great uncertainty remained over how the complex mix of pollution affected the climate.

Scientists had long theorized that air pollution might be "seeding" the formation of clouds. But decades of cloud-seeding experiments had failed to provide proof, and evidence for pollution-related clouds was tenuous. More conclusive evidence came in 1987, when satellite photos revealed persistent clouds over areas of the oceans used as shipping lanes. Smokestack exhaust from ships, dense with sulfate aerosols, was creating clouds that likely reflected sunlight and decreased the solar energy warming the ocean surface.

When Mount Pinatubo in the Philippines erupted, climate scientists seized the opportunity to test their models. The eruption released some 20 million tons of sulfur dioxide into the atmosphere, giving rise to a lingering haze of sulfate aerosols. NASA researchers led by James Hansen calculated that Pinatubo's eruption would lower average global temperatures over the next few years by roughly half a degree Celsius (0.9°F), with the greatest changes in the higher northern latitudes. The prediction proved remarkably on target. By the mid-1990s, most scientists agreed that human-made aerosols were acting like an ongoing volcanic eruption, and that air pollution had likely been masking the impact of global warming for decades.

In a $25 million multinational study spanning four years, climate scientists led by Veerabhadran

Ramanathan documented how pollution was severely dimming areas of the Indian Ocean. The study, called Project INDOEX, found that over northern regions of the ocean, where pollution streams in from India, a pollutant layer nearly two miles thick cut down the sunlight reaching the ocean by more than 10 percent— a far bigger effect than most scientists had thought possible. Ramanathan's own models had led him to expect a dimming of only one half to one percent. Project INDOEX showed in detail how the toxic mix of soot, sulfates, and other pollutants both directly blocked sunlight and, even more critically, helped spawn clouds that reflected the sun's energy back to space.

In the mid-1980s, when meteorologist Gerry Stanhill reported that a dramatic 22 percent reduction of sunlight had occurred in Israel between the 1950s and the 1980s, the news hardly made a splash in the scientific community or popular press. But Stanhill was not alone in measuring such a drop. When he combed the scientific literature, he found that other scientists had measured declines of 9 percent in Antarctica, 10 percent in areas of the U.S., 16 percent in parts of Great Britain, and almost 30 percent in one region of Russia. Alarmed by the trend, Stanhill coined the term "global dimming."

No one is sure what's causing it, or what it means for the future, but it is now reasonably well established that the quality and quantity of the sun's radiation reaching the surface of Earth is being diminished by something, and air pollution in general and contrails in particular are the prime suspects.

Some scientists have reported finding levels of solar radiation reaching the Earth's surface decreasing by almost 3% per decade. This "global dimming" is a growing segment of the global warming controversy.

Global dimming is too small to detect with the naked eye, "but," reports the UK's *Guardian* newspaper, "it has implications for everything from climate change to solar power and even the future sustainability of plant photosynthesis." In any greenhouse, the rule of thumb is that for every 1% decrease in solar radiation the result is a 1% drop in plant productivity. The global warming greenhouse is getting its windows fogged!

AVIATION SMOG

Contrails are the visible part of aviation smog. Aircraft engines daily dump vast amounts of water, CO_2, soot particles and a variety of other chemical substances (which may or may not be part of a global conspiracy) into the atmosphere. Air traffic thus not only alters the properties of the higher regions of our atmosphere but also all the regions below, right down to ground level, and even below that!

Samples from private wells and municipal water supplies have found chemicals from all fuels, particularly MTBEs, in our wells and aquifers. MTBE (methyl tertiary-butyl ether) is a member of a group of chemicals commonly known as fuel oxygenates and is used in automotive fuels (gasoline) to reduce carbon monoxide and ozone levels caused by auto emissions. Daniel P. Jones, Environment Writer for the *Hartford Courant*, reported in 5 January 1999 that:

> As soon as MTBE hit the pumps, in 1992 in some states and in 1995 in much of Connecticut, complaints of health problems began. Motorists said they had trouble breathing, nausea, sore throats, skin rashes, eye irritations and neurological problems after pumping gas or breathing automobile exhaust.
>
> MTBE has been banned in Alaska and part of Montana, because of such complaints. North Carolina banned it after classifying it a probable cause of cancer

in people. Maine is in the process of getting MTBE out of the gasoline sold there. It's under fire in statehouses in other states as well, including California, where some major water supplies have been polluted, and New Jersey, where some of the biggest protests have been staged by motorists who say breathing the fumes makes them sick.

In its fact sheet "MTBE in Drinking Water" the United States Environmental Protection Agency (EPA) admits that "due to its widespread use, reports of MTBE detections in the nation's ground and surface water supplies are increasing" and that one route for it to enter into ground water is through 'air deposition.'" Although MTBE is not used in jet fuels it is a marker that air pollutants are migrating underground.

A lab test done in September 1997 on a sample of jet fuel by Aqua Tech Environmental Labs in Ohio found 51 toxic substances, including ethylene dibromide (EDB). Banned in 1983 by the Environmental Protection Agency in a rare emergency order, EDB is a potent pesticide, chemical irritant and known carcinogen. Military aircraft routinely jettison fuel to lower aircraft weight for safe landings.

Aviation smog is now being increasingly recognized by scientists and concerned global citizens as being responsible for more clouds, rain, drought, bigger hailstones (megacryometeors), and maybe even giving extra power to hurricanes. Still, aviation smog has not received the scientific and political attention many feel it deserves.

Jet aircraft also leave behind unseen carbon dioxide and oxides of sulfur and nitrogen. These emissions have consequences similar to other fossil fuel use. Aviation produces a relatively small amount of these pollutants (13 percent) when compared to other transportation (trains, ships, trucks, cars and such) and a fraction of the global emissions (2 percent) from all sources. Modern jet engines are among the most efficient of all internal combustion engines, thanks in part to the High Bypass Turbofan. Recently the increase in the

247

number air passenger miles being flown has environmental scientists and environmentally minded people concerned.

Surging air travel and transport are pushing fuel consumption steeply upward and by mid-century airplanes may outpace other carbon dioxide sources, especially if countries make good on their promises to limit greenhouse gas emissions under the Kyoto Protocols.

As well as carbon dioxide, planes also emit nitrogen oxide. *Wikipedia* tells us:

> Nitrous oxide, also known as dinitrogen oxide or dinitrogen monoxide, is a chemical compound with chemical formula N_2O. Under room conditions, it is a colourless non-flammable gas, with a pleasant, slightly-sweet odor. It is commonly known as laughing gas due to the exhilarating effects of inhaling it, and because it can cause spontaneous laughter in some people; it is also known as NOS or nitrous in racing and motor sports, where its usage is widespread. It is used in surgery and dentistry for its anaesthetic and analgesic effects. Nitrous oxide is present in the atmosphere where it acts as a powerful greenhouse gas.

Nitrogen oxides stimulate the formation of ozone. Ozone in the stratosphere plays a protective role where it blocks out harmful ultraviolet (UV) radiation from the sun. But close to the ground ozone is a pollutant that endangers the health of humans and plants. To limit this hazard the International Civil Aviation Organization has set standards for nitrogen oxide emissions during takeoff and landing.

Yet the majority of aircraft spend most of their time and release most of their nitrogen oxides at the top of the troposphere, the lowest level of the atmosphere. While ozone produced at cruising altitude is too high to threaten health directly, it is still too low to be of much use in blocking UV, which happens much higher in the stratosphere. Its most important effect then may be as a greenhouse gas, trapping thermal energy and thus possibly contributing to global warming.

Studies using computer models suggest that aircraft nitrogen emissions could have boosted tropospheric ozone concentrations by several percent, especially over the heavily traveled North Atlantic. But a 1994 report by the World Meteorological Organization warned, "little confidence should be put in these quantitative model results of subsonic aircraft effects on the atmosphere."

There are many uncertainties that undermine the reliability of model results. Current models include only some chemical reactions and scientists fear that they may be missing important ones. In addition, researchers do not know what quantity of nitrogen oxides come from other sources, such as lightning. Estimates of lightning's input could be off by several hundred percent, warns Howard L. Wesoky, of NASA headquarters in Washington, D.C.

CONTRAILS AND SCIENTIFIC RESEARCH

Contrail formation, growth, and dissipation and their optical properties (density, reflectivity, etc.) are highly dependent on a number of factors such as: aircraft engine type; and the altitude, temperature, humidity, wind speed and direction of flight. The contrail-cirrus radiative effects (trapping heat or radiating it out into space) likewise depend on several factors including: the underlying conditions (surface temperature and albedo [the amount of light being reflected back off of land and water surfaces]), the optical properties of the contrails themselves, air traffic density and altitude, and the time of day when the contrails are formed, all of which are now the subject of scientific study. Unfortunately, if scientists are also studying the effects of secret additives they aren't mentioning it in public.

Cirrus clouds affect Earth's climate by reflecting incoming sunlight by day and inhibiting heat loss from the surface of the planet by night. It has been estimated that in certain heavy air traffic corridors, cloud cover has increased by as much as 20%. Since contrails can spread out and become cirrus clouds nearly indistinguishable from Mother Nature's own, it is felt by some scientists that contrails may be affecting the planetary climate in similar ways. Many studies are currently underway to better understand both how artificial clouds made by jets alter the climate and the role that jet exhaust aerosols play in modifying the chemistry of the atmosphere.

Contrails have been recorded from the Sahara Desert to the South Pole indicating that contrails are not confined only to the populated regions of Earth. For mainstream scientists contrails are a concern in climate studies as increased jet traffic may have resulted in an increase in global cloud cover, altering the Earth's temperature as a result. Chemtrails, if any of the conspiracy stories are true, may be of even greater concern to us, as they point to dire political and social circumstances in addition to the environmental concerns.

Several scientific studies are now being conducted with respect to contrail formation and their climatic effects. Numerous scientific organizations have taken an interest in them, with new bodies and commissions being formed every year.

The United States Department of Energy (DoE) is heavily involved in this research, with many projects and facilities. For example, one DoE unit is the Atmospheric Radiation Measurement Program (ARM). It was created to help resolve scientific uncertainties related to global climate change, with a specific focus on the role of clouds and their influence on radiative feedback processes (reflecting sunlight/inhibiting heat loss). The primary goal of the ARM Program is to improve the treatment of cloud and radiation physics in global climate models in order to improve the climate simulation capabilities of these models.

Then there's the Carbon Dioxide Information Analysis Center (CDIAC) and its associated World Data Center for Atmospheric Trace Gases. The Climate Change Research

250

Division of the DoE's Office of Biological and Environmental Research supports CDIAC. CDIAC represents DoE in the multi-agency Global Change Data and Information System. They are rather proud of being the primary global-change data and information analysis center of the DoE. Their website says:

> CDIAC responds to data and information requests from users from all over the world who are concerned with the greenhouse effect and global climate change. CDIAC's data holdings include records of the concentrations of carbon dioxide and other radiatively active gases in the atmosphere; the role of the terrestrial biosphere and the oceans in the biogeochemical cycles of greenhouse gases; emissions of carbon dioxide to the atmosphere; long-term climate trends; the effects of elevated carbon dioxide on vegetation; and the vulnerability of coastal areas to rising sea level.

Another American board is the National Academy of Sciences Board on Atmospheric Sciences and Climate. Per their website:

> The Board seeks to advance understanding of the Earth's atmosphere and climate, to help apply this knowledge to benefit the public, and to advise the federal government on issues within the Board's areas of expertise. The Board carries out its mission through the activities of specialized committees and panels charged with providing cogent and independent advice on critical scientific issues, from narrowly defined, highly technical problems to broad public policy concerns.

The National Aeronautics and Space Administration (NASA) has also jumped on this bandwagon, particularly their Earth Science Enterprise Data and Services division.

251

NASA's stated mission is "to understand and protect our home planet, explore the universe and search for life, and inspire the next generation of explorers." The goal of NASA's Earth Science Enterprise (ESE) is:

> ...to use the unique view from space to study, understand, and improve prediction of climate, weather, and natural hazards. NASA's Earth Observing System Data and Information System (EOSDIS) manages and distributes data products via several Distributed Active Archive Centers (DAACs), which all hold data within a different Earth science discipline. Scientists use NASA ESE data to better understand the scope, dynamics and implications of global change within the Earth system.

Scientific interest in clouds and aviation smog is not confined to governmental bodies. For example, two privately owned outfits involved in this research are the Center for Aerosol and Cloud Chemistry and the Center for Atmospheric and Environmental Chemistry at Aerodyne Research, Inc. (ARI) of Billerica, Massachusetts. ARI performs laboratory and field experiments to "understand processes associated with aerosol and cloud particles in the atmosphere and environmental transformations and environmental fate of pollutants and biospheric emissions."

Government agencies, private companies and private sector research consortia support research and development activities at these ARI labs. Government sponsors include: the National Science Foundation; the National Aeronautics and Space Administration; the National Oceanic and Atmospheric Administration; the Department of Energy; the Environmental Protection Agency; the Army Research Office; the Army Corps of Engineers; the Air Force Geophysics Directorate and the Office of Naval Research (that last one, ONR, also funds HAARP). Private sector sponsors include: the Electric Power Research Institute; the Coordinating Research Council; Air

Products, Inc.; the Alternative Fluorocarbon Environmental Acceptability Study; Allied Signal Corp.; Albermarle Chemical Co.; the Gas Research Institute; the Methyl Bromide Global Coalition and the Chemical Manufacturers Association.

Of course this scientific interest extends far beyond the borders of the United States. One example of nascent North American regional government is the *NARSTO* network. Their website explains that *NARSTO* was:

> ...formerly an acronym for "North American Research Strategy for Tropospheric Ozone," the term NARSTO has become simply a wordmark signifying this tri-national, public-private partnership for dealing with multiple features of tropospheric pollution, including ozone and suspended particulate matter. NARSTO is a public/private partnership, whose membership spans government, the utilities, industry, and academe throughout Mexico, the United States, and Canada. Its primary mission is to coordinate and enhance policy-relevant scientific research and assessment of tropospheric pollution behavior; its activities provide input for science-based decision-making and determination of workable, efficient, and effective strategies for local and regional air-pollution management. In accomplishing this mission, NARSTO is charged with establishing and maintaining effective communication channels between its scientific effort and its client community of planners, decision-makers, stakeholders, and strategic analysts. It is also charged with providing a cross-organization planning process, which determines the most effective strategies for scientific investigation. NARSTO coordinates the allocation of financial resources to implement these strategies and monitors progress of its effort toward fulfillment of its programmatic goal. NARSTO was initiated by a formal Charter-signing ceremony at the White House on February 13, 1995.

And of course the UN is in on the action too. One of their many agencies involved in this is the previously mentioned Intergovernmental Panel on Climate Change (IPCC), composed of over 1500 scientists from around the world. The World Meteorological Organization (WMO) and The United Nations Environment Programme (UNEP) established the IPCC with the mission to "assess scientific, technical and socio-economic information relevant for the understanding of climate change, its potential impacts and options for adaptation and mitigation."

The old phrase "too many cooks spoil the soup" comes to mind. I can't help but wonder if all these learned folks and governmental and non-governmental (NGO) organizations might be doing more harm than good, the possibilities of which we will examine in depth in the next chapter.

It also looks to me like there is something of a governmental-academic feeding frenzy taking place. Being the realistic cynic that I am I just can't believe that every individual in all these organizations is motivated solely by concerns over the environment. If nothing else this "global concern" is a grand excuse to shake down the governments of the world, and wring charity from private citizens, to the tune of billions of dollars annually.

What's more, I believe there is something of a political turf war hidden under all this. Many of these organizations are not just competing for grant monies and donations, but are using global warming and other environmental concerns, like access to fresh water, to gain political clout. There has been a movement to create a single world government growing for over 200 years. I see these trans-border problems as just the kind of excuses needed to build a true global government on, as I will expand on in the final chapter.

WHAT HAS NASA FOUND?

For the past decade NASA has held a conference called

the Atmospheric Effects of Aviation Project (AEAP). Several hundred researchers from around the world attend annually. In 1997 Researchers from NASA Langley Research Center in Hampton, Virginia, presented evidence that contrails are contributing to global warming and causing local effects over areas with heavy air traffic.

As I have mentioned, not all scientists buy the "greenhouse gasses cause global warming" argument. Jim Scanlon, a journalist in attendance at the 1997 AEAP conference, reported that the previously mentioned Fred Singer held a session that year that argued that the steady increase in air traffic for the last 20 years was responsible for the nighttime warming detected over North America, not global warming.

Siegfried Frederick Singer (born 27 September 1924 in Vienna, Austria) is a distinguished atmospheric physicist who holds a Ph.D. in physics and is a fellow of the American Physical Society, and is also Distinguished Research Professor at George Mason University and Professor Emeritus of environmental science at the University of Virginia. He is President and founder of the Science & Environmental Policy Project, a non-profit policy research group disputing the greenhouse gas-based global warming and ozone depletion theories. He strongly disagrees with IPCC conclusions about how much warming is to be expected.

On the other side of the coin, the folks at NASA Langley are largely in agreement with the IPCC's conclusions. The Climate Science Branch at NASA's Langley Research Center in Hampton, Virginia, uses "observations from satellite instruments to improve understanding of clouds, aerosols, ozone, and the Earth's radiation balance." Radiation balance is a term for the balance of the solar energy reaching the Earth versus the amount being radiated out into space. Several satellites have been launched into Earth's orbit that indirectly measure the energy absorbed and radiated by the Earth and by inference the energy stored. The Climate Science Branch also uses computer models to simulate cloud processes. Other research activities include "converting satellite data into measurements useful to the renewable energy community and testing the accuracy of satellite measurements through field experiments."

Dr. Patrick Minnis is a Senior Research Scientist with the Climate Science Branch. He is involved in the study of remote sensing of the atmosphere and the Earth's surface with satellites. While many of us have looked at contrails from the ground and wondered what was going on up there, Dr. Minnis has looked down via satellite images and wondered the same thing.

On the evening of 28 July 1998 *NBC Nightly News* reported on Dr. Minnis' work in their story "NASA believes jet contrails contribute to climatic changes." They told how he watched a contrail left behind by a single test flight drift across California for six hours and slowly evolve into a 60-mile-long cloud system. NBC said that on another occasion Minnis watched a figure-8 cloud create its own 60-mile swath of clouds over Texas and Louisiana during a nine-hour period. In 1997, following Hurricane Nora, moist air blanketed the nation's midsection from Nebraska to Texas. Minnis' research team was reported as being stunned by then seeing scores of contrails fuse into one enormous cloud stretching for more than 800 miles.

"We were very excited because it opened our eyes up to the fact that there's possibly a lot of clouds up in the sky that were originally contrails," Minnis said. Just about anyone reading this book could have told him that! For this kind of brilliance we spend billions of taxpayer dollars?

According to Dr. Minnis, "The number of clear days over the U.S. has decreased in the last 30 years, and we suspect that much of that is due to an increase in cirrus clouds, which we suspect is probably due to an increase in air traffic," meaning an increase in contrails—or chemtrails—creating artificial cirrus resulting in dimming.

One question Dr. Minnis addressed was just how often are atmospheric conditions conducive to contrail formation? According to him: "At flight altitudes, conditions that support contrail-generated cirrus exist 10% to 20% of the time in clear air and within standing cirrus"

Although this is a somewhat small percentage, the diverse weather of North America coupled with the staggering number of planes in the air at any given time (globally there are over five million commercial flights per year) results in at least some part of the United States having good contrail making weather on any given day. Worldwide, contrails are estimated to cover .1% of the Earth's surface area and that number is forecast to rise to .5% by 2050.

In 1999 the American Geophysical Union issued a press release with the title "Jet Contrails To Be Significant Climate Factor By 2050" commenting on the work of a research team of American and German scientists, headed by the above-mentioned Dr. Minnis. This PR release stated that:

> By the year 2050, increased flights by jet airplanes will impact global climate through the greater number of contrails they will produce, according to a new study in the July 1 issue of the journal, Geophysical Research Letters. Contrails are ice clouds created by jet engines and are short lived in dry air, but can persist for hours in moist air and become indistinguishable from natural cirrus clouds.

The news release went on to say: "contrails cause a warming of the Earth's atmosphere, although their impact is currently small as compared to other greenhouse effects." Minnis and his team were reported to have predicted that the amount of warming attributable to contrails would grow by a factor of six over the next 50 years.

The press release pointed out that air traffic and therefore contrails are not evenly distributed around the globe, but are concentrated over parts of the United States and Europe. In these areas local warming was found to be up to 35 times the global average. Large contrails can now be observed in satellite imagery, sometimes with hundreds of plumes caught in a single frame. "Although their total global coverage has not yet been determined," the article stated, "it is computed from traffic and weather data to amount to 0.1 percent. In the parts of Europe and eastern North America with the heaviest air traffic, however, contrails currently cover up to 3.8 percent and 5.5 percent of the sky, respectively."

Dr. Minnis and his colleagues were reported as finding that global air traffic rose by over seven percent per year from 1994 to 1997, in terms of passenger miles flown. Similar growth is likely to continue for the foreseeable future. Taking into account such factors as number of flights per day, fuel consumption, and altitudes flown, they concluded that by 2050 "average contrail coverage over Europe will be four times higher than at present, or about 4.6 percent. In the United States, the increase will be 2.6 times current levels, or 3.7 percent coverage; and in Asia, the increase will be ten times current levels, or 1.2 percent."

NASA has posted several pieces on Minnis' work to their website. One under the heading of "Clouds Caused By Aircraft Exhaust May Warm The U.S. Climate" leads off with: "NASA scientists have found that cirrus clouds, formed by contrails from aircraft engine exhaust, are capable of increasing average surface temperatures enough to account for a warming trend in the United States that occurred between 1975 and 1994."

Minnis was quoted as saying: "This result shows the increased cirrus coverage, attributable to air traffic, could account for nearly all of the warming observed over the United States for nearly 20 years starting in 1975." Possibly realizing that he was supporting Professor Singer's position and goring the global warming sacred cow that NASA worships, he added: "... but it is important to acknowledge contrails would add to and not replace any greenhouse gas effect. During the same period, warming occurred in many other areas where cirrus coverage decreased or remained steady." Back-pedaling at its finest, eh?

He is further quoted as having said: "This study demonstrates that human activity has a visible and significant impact on cloud cover and, therefore, on climate. It indicates that contrails should be included in climate change scenarios."

Using published results from the general circulation model at NASA's Goddard Institute for Space Studies (New York), Minnis and his colleagues estimated that contrails and their resulting cirrus clouds would have increased surface and lower atmospheric temperatures by 0.36 to 0.54 degrees Fahrenheit per decade. Their calculations were confirmed when weather service data revealed that surface and lower atmospheric temperatures across North America had indeed risen by almost 0.5 degrees Fahrenheit per decade between 1975 and 1994.

NASA reported that both air traffic and cirrus coverage increased during the studied period of warming despite no changes recorded by the National Centers for Environmental Prediction (NCEP) in the humidity at jet cruise altitudes over the United States in the same time period. By contrast, they reported that humidity at flight altitudes decreased over other land areas, such as Asia, and was accompanied by less cirrus coverage, except over Western Europe, where air traffic is very heavy. But if the air is getting warmer it should be holding more moisture. And where is the reported dimming over China coming from, if not clouds?

Cirrus coverage did rise in the North Pacific and North Atlantic flight corridors, NASA reported. The trends in cirrus cover and warming over the United States were greatest during winter and spring, the same seasons when contrails are most frequent. These results, along with findings from earlier studies, led NASA to the conclusion that contrails caused the increase in cirrus clouds.

"This study indicates that contrails already have substantial regional effects where air traffic is heavy, such as over the United States. As air travel continues growing in other areas, the impact could become globally significant," Minnis said.

USA TODAY also reported on the research of Minnis and

his colleagues in "Plane Trails in Sky Turn Up the Heat Below, Study Suggests." In the 29 April 2004 edition Traci Watson wrote that this NASA study found the average temperature to have increased by 1°F. Although a single degree may seem trivial, the incredibly large scale that it applies to makes it significant since just 9°F separates our current average temperature from the last Ice Age!

In the spring of 2006 NASA's Goddard Institute for Space Studies released findings that 2005 was the hottest year on record. "The average global surface temperature of 14.77 degrees Celsius (58.6 degrees Fahrenheit) was the highest since recordkeeping began in 1880," they reported. January, April, September, and October of 2005 were the hottest of those months on record, while March, June, and November were the second warmest ever. A few weeks later the newswires of North America hummed with the word that January 2006 had been the warmest January on record.

The newest numbers from NASA say that during the past century temperatures rose 0.8 degrees Celsius (1.44 degrees Fahrenheit), 0.6 degrees of which occurred during the last three decades. The average temperature of 14.02 degrees Celsius in the 1970s rose to 14.26 degrees in the 1980s; in the 1990s it reached 14.40 degrees Celsius; and during the first years of the 21st Century global temperature has averaged 14.62 degrees Celsius.

Keeping in mind that no one knows for sure why the Earth is getting hotter, NASA can prove that it is, and the heat is increasing. Of the hottest years on record, six of them occurred in the eight years between 1998 and 2005, inclusive. "After 2005, 1998 was the second warmest, with an average global temperature of 14.71 degrees Celsius," NASA reported. They did note that there was an important difference between 1998 and 2005 — the strongest El Niño of the past 100 years lifted the average 1998 temperature by 0.2 degrees Celsius, whereas the record warmth for 2005 was not buoyed by such an effect. Some ideas beyond global warming for where this heating came from have been floated, like increased solar energy

260

reaching the Earth from the recent "solar maximum."

Every 11 years the sun undergoes a period of activity called the "solar maximum," followed by a period of quiet known as the "solar minimum." During the "solar max" there are many sunspots with solar flares erupting near them on a daily basis. Coronal mass ejections, billion-ton clouds of magnetized gas, fly away from the Sun and buffet the planets, all of which can affect communications and weather here on Earth. Even the Sun's magnetic field—as large as the solar system itself—grows unstable and flips. The most recent Solar Max was also the most powerful ever recorded. It, like the previous two solar maximums, had two peak periods, separated by about 18 months. The first peak crested in mid-2000, with the second peak hitting late in 2001. NASA is now predicting that the next Solar Max, due about 2012 (a date prophesied by some [based on the Mayan calendar] to be the End of the World!) will be one and a half times as big as this last one was!

Of course some of the other ideas being floated are a little hard to swallow, like claims that a rogue planet called Planet "X" (or *Nibiru* by some authors, such as Zecharia Sitchin) in an elliptical orbit is about to do a fly-by of the Earth and that is what is destabilizing the Sun and adding energy to the Earth.

CONTRAILS AND 9/11

As you may have guessed, the idea that contrails increase cloudiness is hardly news, even in scientific circles. In 1981, for example, climatologist Stanley A. Changnon of the Illinois State Water Survey in Champaign reported that the Midwest had grown significantly cloudier during the 1960s and 1970s, with the greatest changes seen in areas of high jet traffic. He also noted a narrowing of the gap between high and low temperatures, possibly attributable to the increase in clouds.

More recently, Kuo-Nan Liou, an atmospheric physicist

at the University of Utah in Salt Lake City, examined changes in high clouds. He found a 5 to 10 percent increase in cirrus cover over Salt Lake City, Denver, Chicago, St. Louis, and several other cities between 1948 and 1984. "Statistically, the high-level clouds appear to be increasing. So we speculate that there might be some potential relationship between aircraft activities and these high-level cloud increases," says Liou.

Nick Onkow is a pilot, flight instructor, and a photographer whose photographs can be found at airliners.net. One can also find there his excellent and comprehensive article on contrails and their environmental effects: "Contrails: What's Left Behind Is Bad News." In it he tells us:

> In one study conducted by meteorologist Keith P. Shine, data from satellites was used to prove that only one percent of the increase in clouds throughout the world have been from aircraft. There are also inherent flaws in some of the research performed by NASA. One problem is the difficulty that scientists have distinguishing a suspected contrail cloud from a natural cirrus cloud in satellite images. Skeptics of the theory that contrails do not have an impact on weather argued this theory with some success until a significant event occurred in North America, the main testing grounds of contrail research.
>
> The terrorist attacks of September 11, 2001 was the aforementioned event, and it was likely to have excited meteorological researchers involved in contrail impact studies. The national airspace was shut down for three days, something that had not yet occurred since the jet age began in the 1960s and is not likely to occur ever again. Scientists took advantage of this unique three day period in history that lacked contrails. What they learned was shocking and is enough evidence to effectively silence any counterargument to their case. One measure of climate is the average daily temperature range (DTR). For thirty years this had been recorded

262

and extra cirrus clouds in the atmosphere would reduce this range by trapping heat. "September 11 – 14, 2001 had the biggest diurnal temperature range of any three-day period in the past 30 years," said Andrew M. Carleton. Not in three decades had there been such a large temperature spread between the daytime highs and the nighttime lows. Furthermore, the increase in DTR during those three days was more than double the national average for regions of the United States where contrail coverage was previously known to be most abundant, such as the Midwest, northeast, and northwest regions. The specific increase in the range was 2°F, which in three days was twice the amount the average temperature had increased by over thirty years time. This is evidence that contrails do alter the climate of the land they drift above.

Wired magazine ran "Hot on the Contrails of Weather" by Mark K. Anderson on 15 May 2002, which also reported on these findings. "It has emerged," Anderson wrote, "that the American climate was indeed noticeably different during those three days without air travel."

This article reported on the findings of a team of climatologists who had presented their findings to the American Meteorological Society conference in Portland, Oregon, the week before. Their data addressed variability between day and night temperatures, not on overall warming or cooling trends. They had conclusive proof that temperatures in the United States had fluctuated more when airplanes were grounded than when normal flight patterns had prevailed. Planes in the sky, they said, reduce the difference between day and nighttime temperatures. "More air travel," Anderson wrote, "brings less meteorological difference between noon and midnight."

"We actually found a much greater change in temperature range for parts of the country that normally get the greatest contrail coverage," said David J. Travis, of the University of

Wisconsin in Whitewater.

Over nearly all of the country they found this effect, but in normally contrail-heavy portions of the country the difference between daytime and nighttime temperatures had changed dramatically. The Midwest and Northeast in particular had experienced a "contrail effect" of 3 degrees Celsius, more than twice the national average. Proponents of global warming theories believe it only takes climatic changes of fractions of a degree Celsius to yield widespread results.

As air traffic increases over some regions of the world, the increased density of contrails will likely bring even smaller differences between daytime and nighttime temperatures, and that, Anderson wrote, will alter the local environment. For instance, he pointed out that cranberry bogs and citrus orchards require a combination of cool nights and warm days for optimum yield. And in the spring, sugar maples don't produce sap if daily (diurnal) temperatures don't fluctuate enough. Furthermore, some insects are particularly sensitive to changes in diurnal variations. And changes in insect populations can in turn have some unexpected consequences.

The previously mentioned Dr. Patrick Minnis of NASA's Langley Research Center said Travis' results confirm previous statistical studies published on climate variability and contrails. "Having this data set made the relationship a little more definitive," Minnis said.

In that same week Minnis had presented his research on contrails using the unique window that the grounding of aircraft in the aftermath of the September 11 attacks provided. However, instead of studying the lack of airborne jets during the FAA's three-day moratorium, Minnis considered the few aircraft that were in the skies—military jets and transport planes (I wonder if he noticed the 747s full of Saudis scurrying home?).

Anderson reported that in the usually packed air corridor around Washington, D.C., Minnis was able, using satellite images, to follow a lone contrail drifting through the mid-

Atlantic states on 12 September. The three days of grounded air travel provided Minnis with a unique opportunity to model the evolution of that single contrail where normally scores or even hundreds would usually be found. In all Minnis tracked six contrails, each no wider than an airplane wing, as they evolved in a matter of hours into cloudbanks that covered 20,000 square kilometers.

"This is a once-in-a-lifetime opportunity to measure these contrail effects," Travis said. "Or, at least, we can only hope its once in a lifetime."

So, we have seen that jet engines have changed, increasing the likelihood of persistent contrails. And we have seen that the atmosphere itself has been changed in some fundamental way by a cause or causes as yet unknown. Contrails, particularly persistent contrails, seem to be significant contributors to both global warming and global dimming. Yet, humidity at flight altitudes has not increased, according to some studies. If increased moisture isn't creating contrails, what is? We have seen that aviation fuel contributes to aviation smog, which may account for some of these effects. But what if it's more? What about fuel *additives* or the intentional release of other aerosols? We will take up this question in the next chapter.

CHAPTER SEVEN

GEOENGINEERING

If you go looking for information on contrails/chemtrails on the Internet you have to know the key words to search with. "Chemtrails" will pull up sites dealing with the conspiracy, but you will find little or no corroborating government or scientific sites. "Aerosols" is a little better. It will get you a lot of scientific sites involved in monitoring greenhouse gases and other airborne pollutants. For the "mother lode" try *Googling* "Geoengineering."

The Encyclopedia of Global Change defines Geoengineering as:

> ...intentional large scale manipulation of the global environment. ... Geoengineering schemes seek to mitigate the effect of fossil-fuel combustion on the climate without abating fossil fuel use; for example by placing shields in space to reduce the sunlight incident on the Earth.

Climate Change 2001, a report from the previously mentioned Intergovernmental Panel on Climate Change, confirms that geoengineering:

> ...includes the possibility of engineering the Earth's climate system by large-scale manipulation of the global energy balance. It has been estimated, for example, that the mean effect on the Earth surface energy balance from a doubling of CO_2 could be offset by an increase of 1.5% to 2% in the Earth's albedo, i.e. by reflecting additional incoming solar radiation back into space.

A SUNSCREEN FOR THE PLANET?

One of the most widely discussed, and plausible, reasons suggested for the spraying of chemtrails is mitigation (corrective measures) taken to offset the effects of global warming. In 1979, famed physicist Freeman Dyson proposed that the deliberate, large-scale introduction of fine particles (aerosols) into the upper atmosphere would offset global warming. Dr. Dyson's concept was fairly simple. The Earth's albedo is the amount of sunlight reflected by our planet back into space. If you increase the albedo you reduce the amount of sunlight and warmth that would reach the Earth's surface, causing a reduction in the amount of heating taking place. If these greenhouse gases really were causing a rise in temperatures, it was reasoned that this rise could be cancelled out by increasing the Earth's albedo, i.e., by intentional global dimming.

One can get a feel for how much sunlight is reflected in the albedo by looking at a new moon. On those nights when you can see only a tiny sliver of the moon, you can sometimes also just make out the dark disk of the moon beside the brightly shining crescent. That dark portion of the moon is being illuminated by sunlight reflected off the Earth and then bounced back off the moon to your eyes.

Mother Nature routinely changes the Earth's albedo with clouds of ash spewed from volcanoes. It is well documented that the vast amounts of ash and pulverized rock injected into the upper atmosphere by a volcanic eruption can affect the weather for months, even years after the event. Charles Pellegrino, writing in *Unearthing Atlantis* about the cataclysmic eruption of Thera (the Greek island of Santorini), likened it to 1816, "the year without a summer," that followed the explosive reawakening of the Tambora volcano in Indonesia. He wrote:

> There were no harvests in New England that year [1816]. Megatons of ultra fine dust had been hoisted

fifty miles high into the stratosphere, where it shaded out some of the sun's radiation, absorbing its heat long before it reached the ground. As June and August snowstorms swept across New York, few people could draw consolation from the strange beauty of a blood red moon, or from the most splendid sunsets the world had seen in more than thirty-four hundred years.

Dr. Edward Teller, "the father of the hydrogen bomb," with Dr. Lowell Wood who was of both the Lawrence Livermore National Laboratory (LLNL) and the Hoover Institution on War, Revolution and Peace at Stanford University, plus some of their colleagues from LLNL surveyed the technological prospects for implementing Dr. Dyson's idea. They estimated the costs involved and presented their results in a paper titled *Global Warming and Ice Ages: Prospects for Physics-Based Modulation of Global Change* which was prepared for invited presentation at the Twenty-Second International Seminar on Planetary Emergencies in Erica, Italy, 20-23 August 1997. They concluded that Dyson's "geoengineering" scheme could cost as much as $1 billion a year. However, Teller's team also thought that a more technologically advanced option along the same lines might cost as little as $100 million.

Teller then wrote an article touting the proposal for The *Wall Street Journal*, which ran it on 17 October 1997. Teller's piece was titled "The Planet Needs a Sunscreen." It was written in a flip, tongue-in-cheek style that presented the whole concept of global warming as a laughable non-issue. But, were global warming a fact, which he clearly doubted, one approach to dealing with it that he found particularly appealing would be to implement Dyson's concept of increasing the Earth's albedo, which Teller jocularly referred to as a "sunscreen."

Teller suggested that slightly diminishing the amount of sunlight reaching the Earth's surface (perhaps by as little as just 1%) would offset the alleged warming effect of greenhouse gases for decades, perhaps even centuries to come. In the article Teller off-handedly commented that the Director of the

U.S. Global Change Research Program's Coordination Office had been promoting such a geoengineering scheme for three decades!

As they say, he should know, for his people had been taking money from the US government for this very research for a decade or more! Just one of the many programs at LLNL that Teller and Wood's paper was an outgrowth of is The Program for Climate Model Diagnosis and Intercomparison (PCMDI). The PCMDI was established in 1989 at the LLNL and funded by the Climate Change Research Division of the U.S. Department of Energy's Office of Science, Biological and Environmental Research (BER). The PCMDI's mission is to develop "improved methods and tools for the diagnosis, validation, and intercomparison of global climate models, and to conduct research on a variety of problems in climate modeling and analysis."

The Committee on Science, Engineering, and Public Policy (COSEPUP) is a joint unit of the National Academy of Sciences, National Academy of Engineering, and the Institute of Medicine. In a paper they presented in 1992 titled: *Policy Implications of Greenhouse Warming: Mitigation, Adaptation, and the Science Base*, they analyzed several systems for mitigating global warming. Four of the remedies considered involved different ways of injecting "dust" (aluminum oxide) into the atmosphere to change the Earth's albedo. One system proposed using a shipboard naval rifle system to fire millions of canisters of dust into the sky over a 40-year period. Two of the proposals involved using balloons to deliver and release the dust high in the atmosphere, and the fourth specifically called for increasing cloud abundance through creating aluminum oxide chemtrails! The report said:

> Several schemes depend on the effect of additional dust compounds in the stratosphere or very low stratosphere screening out sunlight. Such dust might be delivered to the stratosphere by various means, including being fired with large rifles or rockets or

270

being lifted by hydrogen or hot-air balloons. These possibilities appear feasible, economical, and capable of mitigating the effect of as much CO_2 equivalent per year as we care to pay for. Such systems could probably be put into full effect within a year or two of a decision to do so, and mitigation effects would begin immediately. Because dust falls out naturally, if the delivery of dust were stopped, mitigation effects would cease within about 6 months for dust (or soot) delivered to the tropopause and within a couple of years for dust delivered to the midstratosphere.

Perhaps one of the surprises of this analysis is the relatively low costs at which some of the geoengineering options might be implemented.

Policy Implications of Greenhouse Warming was a scientific report on greenhouse gases, global warming, policy decisions and mitigations. Included are the scientists, agencies, institutions and corporations involved, cost factors, chemical formulas, mathematical modeling, delivery methods, policies, recruiting of foreign governments, acquisition of materials, and the manufacturing of aerosol compounds, etc. At the time of this writing the entire 994-page study can be read online at: www.books.nap.edu/books/0309043867/html/index.html/

The conclusion reached by the authors of *Policy Implications of Greenhouse Warming: Mitigation, Adaptation, and the Science Base* was that the most effective global warming mitigation alternative investigated by their committee turned out to be the spraying of reflective aerosol compounds into the atmosphere utilizing commercial, military and private aircraft — chemtrails!

This preferred mitigation method could be utilized to both create a global atmospheric shield by increasing the Earth's albedo using aerosol compounds of aluminum and barium oxides, and could at the same time introduce ozone generating chemicals into the atmosphere, replenishing the ozone layer. They concluded that this method was the most

271

cost effective, and thus yield the largest benefits. It could also be conducted covertly to avoid the burdens of environmental protection and regulatory entanglements!

Another revealing paper is "Geoengineering: A Climate Change Manhattan Project" by Jay Michaelson, published in the *Stanford Environmental Law Journal*, January 1998. Michaelson makes a case for the pressing need for undertaking geoengineering projects immediately. He also argued that regulation, environmental laws and other stumbling blocks limit our ability to address the climate change dangers that he believed threaten us. To save us from ourselves he proposed a project as large and as secret as the Manhattan Project that developed America's first nuclear weapons during World War II. He wrote:

> The projected insufficiency of Kyoto's emission reduction regime, and the problems of absence, cost, and incentives discussed in part II, cry out for an alternative to our present state of climate change policy myopia. Geoengineering — intentional, human-directed manipulation of the Earth's climatic systems — may be such an alternative. This part proposes that, unlike a regulatory "Marshall Plan" of costly emissions reductions, technology subsidies, and other mitigation measures, a non-regulatory "Manhattan Project" geared toward developing feasible geoengineering remedies for climate change can meaningfully close the gaps in global warming and avert many of its most dire consequences.
>
> In some ways, this phase has already begun, as geoengineering has moved from the pages of science fiction to respectable scientific and policy journals. One of the most encouraging proposals today focuses on the creation of vast carbon sinks by artificially stimulating phytoplankton growth with iron "fertilizer" in parts of the Earth's oceans. Another proposal suggests creating miniature artificial "Mount Pinatubos" by allowing

airplanes to release dust particles into the upper atmosphere, simulating the greenhouse- arresting eruption of Mount Pinatubo in 1991.

Michael Behar's article "How Earth-Scale Engineering Can Save the Planet" in the August 2005 issue of *Popular Science* described a meeting at the White House in September 2001 organized by President George W. Bush's Climate Change Technology Program to discuss *Response Options to Rapid or Severe Climate Change*. "While administration officials were insisting in public that there was no firm proof that the planet was warming," Behar reported that in private "they were quietly exploring potential ways to turn down the heat."

Shortly after assuming office President George W. Bush outraged Greens around the world by officially withdrawing US support from the Kyoto Protocol. This meeting therefore represented something of a US counterproposal to Kyoto — one that included variations on Edward Teller's "Sunscreen for Planet Earth"!

Physicist and economist David Keith was one of the more than two dozen scientists — including physicists from Teller's LLNL who had spent much of their careers designing nuclear weapons — who attended that White House conference. Dr. Keith was quoted in the article as saying: "If they had broadcast that meeting live to people in Europe, there would have been riots. Here were the bomb guys from Livermore talking about stuff that strikes most greens as being completely wrong and off-the-wall."

The *Popular Science* article went on to report that a growing number of physicists, oceanographers and climatologists around the world were seriously considering technologies for the deliberate manipulation of Earth's climate. These plans included orbiting space mirrors to deflect sunlight away from Earth, and using unmanned sailboats controlled by computers zigzagging back and forth across the oceans while using solar power to generate fog that would create a reflective layer of clouds over the Earth's seas.

273

Edward Teller concluded his *Wall Street Journal* article, "The Planet Needs a Sunscreen," with:

> Yet if the politics of global warming require that "something must be done" while we still don't know whether anything really needs to be done — let alone what exactly — let us play to our uniquely American strengths in innovation and technology to offset any global warming by the least costly means possible. While scientists continue research into any global climatic effects of greenhouse gases, we ought to study ways to offset any possible ill effects.
>
> Injecting sunlight-scattering particles into the stratosphere appears to be a promising approach. Why not do that?

DR. STRANGELOVE

Edward Teller (15 January 1908 – 9 September 2003) was a Hungarian-born American nuclear physicist. He was a co-founder of the Lawrence Livermore National Laboratory (LLNL) and was both its director and associate director for many years. Before that he had been a member of the Manhattan Project. David Shukman wrote in *Tomorrow's War: The Threat of High-Technology Weapons* that during the Manhattan Project:

> Teller sulked and left when Oppenheimer refused him a key post; later, when the two clashed over whether the United States should develop the H-bomb, Teller implied to a congressional committee that Oppenheimer was a security risk, testimony which led to the latter losing his security clearances and to his subsequent ostracism from the nuclear community.
>
> After his testimony Teller was treated as a pariah by many of his former colleagues. Teller in turn shunned liberal academics to become the scientific darling of

conservative politicians and military strategists for his advocacy of American scientific and technological supremacy.

Wikipedia tells us that:

> Teller's vigorous advocacy for strength through nuclear weapons, especially when so many of his wartime colleagues later expressed regret about the arms race, made him an easy target for the "mad scientist" stereotype (his accent and imposing eyebrows certainly did not help shake the image).

Many people believe that Teller was the inspiration for the character of Dr. Strangelove in Stanley Kubrick's 1964 satirical film *Dr. Strangelove: Or How I Learned To Stop Worrying And Love The Bomb.*

Sponsored by the scientific humor journal *Annals of Improbable Research* the Ig Nobel Prizes are a parody of the Nobel Prizes. The name is a play on the words ignoble and "Nobel" and are given for discoveries "that cannot, or should not, be reproduced." In 1991 Teller was awarded one of the first ever Ig Nobel Prizes for Peace in recognition of his "lifelong efforts to change the meaning of peace as we know it."

Teller's courting of favor from the military and governmental agencies left him in a position to do far more than just write far-fetched articles for Dow Jones—could he have actually been fronting for someone or some agency that was, or soon would be spraying us? These chemtrails sound like science gone mad—could they be the work of real-life Dr. Strangeloves?

STRATOSPHERIC WELSBACH SEEDING

Further evidence to support the belief that chemtrails

are indeed the result of such a geoengineering scheme is found in a United States Patent (number 5,003,186) that was granted to David B. Chang and I-Fu Shih of the Hughes Aircraft Company in 1991 for *Stratospheric Welsbach Seeding For Reduction Of Global Warming*.

Auer von Welsbach discovered what is called the Welsbach effect, which is the basis of an incandescent gas burner invented by him. The ubiquitous Coleman lantern is the most popular version of Welsbach's invention. It works by burning a mixture of air and gas or vapor to heat a mantle to incandescence. The mantle is made by soaking a "stocking" in a solution of nitrates of thorium and cerium and, for use, igniting it to burn the thread, converting the nitrates into oxides, which remain as a fragile ash. The light far exceeds that obtained from the same amount of gas burned in an ordinary fishtail burner.

Welsbach materials have the characteristic of wavelength-dependent emissivity (the ability to emit or radiate energy, such as heat or light). For example, thorium oxide has high emissivities in the visible and far infrared (IR) regions (light) but it has low emissivity in the near IR region (heat). This characteristic of emitting or reflecting heat or light at some wavelengths but not others is known as the Welsbach effect, as seen at work in the Coleman lantern.

The most commonly proposed materials for this seeding are barium and aluminum. As discussed earlier, barium is hazardous to human health. So is aluminum, which is toxic to the nervous system and deleterious to the brain. Its effects on human health were illustrated when Canadians in Espanola, Ontario, reported mass illness after low-flying U.S. jets "straffed" their town in a military exercise, releasing aluminum coated fibers (chaff) in the spring of 1998. Investigators found area rainwater contained seven times the allowable limit for aluminum exposure, as people complained of neck pain, breathing problems, headaches, burning eyes and dry coughs.

The invention described in the patent for *Stratospheric*

Welsbach Seeding For Reduction Of Global Warming calls for the layer of greenhouse gases to be seeded with Welsbach or Welsbach-like materials. These materials would in effect do the exact opposite of what the greenhouse gases are believed to be doing.

Greenhouse gases are relatively transparent to sunshine but absorb strongly the long-wavelength infrared radiation released by the Earth, trapping the heat in the upper atmosphere. Most current approaches to reduce global warming are to restrict the release of these gases as seen in the Kyoto Protocols. These strategies call for the establishment of regulations on how much of these gases can be produced, where and by whom, and the need to monitor the various gases and to enforce the regulations. Inevitably this would give rise to a global bureaucracy with police powers—a true world government—something very much feared by many people of many different political persuasions.

But what if you could increase the atmosphere's ability to radiate into space the heat held in these gases? If you could send this pent-up heat out into space, we could continue with our polluting technologies and avoid creating a global Big Brother bureaucracy with an environmental agenda. We would still have the gases in the atmosphere but there would be no greenhouse effect to blame for global warming.

The authors of this patent mention Dr. Dyson's idea of increasing the Earth's albedo (seven years before Teller called it a sunscreen for the planet) in what has become the classic chemtrail scenario saying:

> One proposed solution to the problem of global warming involves the seeding of the atmosphere with metallic particles. One technique proposed to seed the metallic particles was to add the tiny particles to the fuel of jet airliners, so that the particles would be emitted from the jet engine exhaust while the airliner was at its cruising altitude. While this method would increase the reflection of visible light incident from space, the

metallic particles would trap the long wavelength blackbody radiation released from the Earth. This could result in net increase in global warming.

Note that last sentence, as this could be critical.

In 1996, Scientists for Global Responsibility reached the same conclusion. This group is associated with the School of Environmental Sciences at the University of East Anglia in Norwich in the United Kingdom. They produced a paper titled: "Climate Engineering: A Critical Review of Proposals." Their report contended that dangerous geoengineering as proposed by Teller et al. would be absolutely ineffective in mitigating global warming for it would have the opposite effect! The report further noted that this climate engineering research was funded by industries with vested interests in the continued high consumption of fossil fuels.

Some of the scientists who have analyzed these chemtrail proposals say that adding aluminum oxide to jet fuel will create more abundant clouds, but because the clouds are below the greenhouse gas layer they will only succeed in creating more heat! Could it be that the fools are trying to "fix" global warming and are actually creating it instead of alleviating it? Or worse, could they be intentionally creating global warming as a pretext to launch their version of a global government whose avowed purpose is to fix global warming (that they are surreptitiously creating)?

Although the method described in the patent for *Stratospheric Welsbach Seeding For Reduction Of Global Warming* does call for seeding (i.e., spraying) quantities of tiny particles (such as thorium or aluminum oxide) into the atmosphere, their plan is to inject them higher than air traffic normally flies.

The atmosphere is divided into three main layers: the troposphere, the stratosphere and the ionosphere. Most commercial flights are short haul "hops" in the troposphere about 4 or 5 miles up. But the greenhouse gas layer typically extends from the top of the troposphere well into the

278

stratosphere between about 6 and 12 miles above the Earth's surface. The authors of the patent proclaim:

> The particles suspended in the stratosphere as a result of the seeding provide a mechanism for converting the blackbody radiation emitted by the Earth at near infrared wavelengths into radiation in the visible and far infrared wavelength so that this heat energy may be reradiated out into space, thereby reducing the global warming due to the greenhouse effect.

In January of 2001, *CBS News* ran a two part *Eye on America Report: Cooling the Planet* that revealed that scientists were "looking at drastic solutions for global warming, including manipulating the atmosphere on a massive scale." The CBS report revealed on air that the global warming reduction methods championed by physicist Edward Teller were being discussed, confirming that a scheme to load the air with tiny particles to "deflect enough sunlight to trigger global cooling" was being planned!

Margareta-Erminia Cassani of moonbowmedia.com posted "CBS News Confirms Global Warming Experiments Underway," an article she wrote on the CBS Report, to Rense. com a few days after the show aired. She quoted Ken Caldeira, the climate researcher from LLNL I quoted from in the last chapter, as saying: "The simplest solution is to put into the high atmosphere small particles which scatter away 1 or 2% of the sunlight...the sooner the better."

She wrote that Caldeira…

> …initially set out to prove Teller's sunlight scattering hypothesis unworkable but when he applied the variables to his geoengineering computer program he not only proved Teller's remedy workable but the best solution possible. Caldeira further opined that another way to

279

accomplish Teller's objectives would be to put a satellite into space between the Earth and the Sun which would activate a 1200 mile solar shield as a sunscreen to block the sun's UV rays. The result would be returning the Earth's climate back to the temperatures that existed before human interference in climate via excessive fossil fuel burning and the uncontrolled use of CFC's.

She also reported that CBS told its audience that there were other methods under experimental investigation at Stanford University to fight global warming. A few of the methods she listed as being highlighted by the CBS report included:

1. Particle blasting. Thousands of light reflective particles would be blasted from warship guns out on the oceans into the skies. The drawback to this method is that blue skies, as we know them, would be a thing of the past. Because the particle blasting would need to occur continuously in order to be effective, blue skies would become mostly a permanent white.

2. Mirrors in space. Another experimental method under consideration is to scatter 50,000 mirrors into space to deflect the sun's rays backward into space, away from the Earth. The drawback of this method is the flickering sun effect that would occur on Earth.

3. Help from the Sea. Still another method involves putting powdered iron into the sea which stimulates the growth of underwater plankton, or algae, which works to absorb more UV rays. The drawback of this is that overgrowths of algae can cause ecosystem upsets in coral systems and other sea life through algal based diseases and/or feeding problems.

She ended her piece with:

Steve Schneider, a Stanford University global warming expert explained that humans don't have 200 years to wait for the Earth to fix the global warming problem itself. We have to do something, now, Schneider said, to help it along or face a multiple of climate changes that can have disastrous effects on man. What that something is remains an experiment at this stage.

Many others in government and the scientific community are taking Teller's sunscreen idea seriously. *Science I Essential Interactions*, published by Centre Point Learning, Inc. of Fairfield, Ohio is a secondary school (7th grade) science textbook now used in some public schools. It enthusiastically touts Teller's sunscreen project by showing a large orange-red jet with the caption, "Jet engines running on richer fuel would add particles to the atmosphere to create a sunscreen." Of course the question remains, are they doing it?

PATENTLY OBVIOUS

United States Patent number 5,003,186 granted to David B. Chang and I-Fu Shih of the Hughes Aircraft Company in 1991 for *Stratospheric Welsbach Seeding For Reduction Of Global Warming* is far from being the only one of its kind. Here is a short list of similar patents granted by the United States Patent Office for this and related technologies.

Patent Number – Date Granted – Title:

3274035 - September 20, 1966 - Metallic Composition For Production Of Hydroscopic Smoke
3518670 - June 30, 1970 - Artificial Ion Cloud
3608820 - September 20, 1971 - Treatment Of Atmospheric Conditions By Intermittent Dispensing Of Materials Therein

3630950 - December 28, 1971 - Combustible Compositions For Generating Aerosols, Particularly Suitable For Cloud Modification And Weather Control And Aerosolization Process USRE29142 - This Patent Is A Reissue Of Patent US3630950 - Combustible Compositions For Generating Aerosols, Particularly Suitable For Cloud Modification And Weather Control And Aerosolization Process

3659785 - December 8, 1971 - Weather Modification Utilizing Microencapsulated Material

3677840 - July 18, 1972 - Pyrotechnics Comprising Oxide Of Silver For Weather Modification Use

3769107 - October 30, 1973 - Pyrotechnic Composition For Generating Lead Based Smoke

3813875 - June 4, 1974 - Rocket Having Barium Release System To Create Ion Clouds In The Upper Atmosphere

3899144 - August 12, 1975 - Powder Contrail Generation

3994437 - November 30, 1976 - Broadcast Dissemination Of Trace Quantities Of Biologically Active Chemicals

RE29,142 - February 22, 1977 - Reissue Of: 03630950 - Combustible Compositions For Generating Aerosols, Particularly Suitable For Cloud Modification And Weather Control And Aerosolization Process

4129252 - December 12, 1978 - Method And Apparatus For Production Of Seeding Materials

4633714 - January 6, 1987 - Aerosol Particle Charge And Size Analyzer

4684063 - August 4, 1987 - Particulates Generation And Removal

4686605 - August 11, 1987 - Method And Apparatus For Altering A Region In The Earth's Atmosphere, Ionosphere, And/Or Magnetosphere [*This, by the way, was the first of the twelve HAARP patents granted.*]

4704942 - November 10, 1987 - Charged Aerosol

4712155 - December 8, 1987 - Method And Apparatus For Creating An Artificial Electron Cyclotron Heating Region Of Plasma

282

4829838 - May 16, 1989 - Method And Apparatus For The Measurement Of The Size Of Particles Entrained In A Gas

4873928 - October 17, 1989 - Nuclear-Sized Explosions Without Radiation

4948050 - August 14, 1990 - Liquid Atomizing Apparatus For Aerial Spraying

4999637 - March 12, 1991 - Creation Of Artificial Ionization Clouds Above The Earth

5038664 - August 13, 1991 - Method For Producing A Shell Of Relativistic Particles At An Altitude Above The Earths Surface

5041760 - August 20, 1991 - Method And Apparatus For Generating And Utilizing A Compound Plasma Configuration

5041834 - August 20, 1991 - Artificial Ionospheric Mirror Composed Of A Plasma Layer Which Can Be Tilted

5104069 - April 14, 1992 - Apparatus And Method For Ejecting Matter From An Aircraft

5912396 - June 15, 1999 - System And Method For Remediation Of Selected Atmospheric Conditions

6030506 - February 29, 2000 - Preparation Of Independently Generated Highly Reactive Chemical Species

6263744 - July 24, 2001 - Automated Mobility-Classified-Aerosol Detector

As mentioned earlier, patents for modifying the weather go back to the 1890s. See Appendix C for a list of approximately 150 such patents, starting in the 1920s.

It would indeed seem that as former Secretary of Defense Cohen said: "There are plenty of ingenious minds out there at work"!

We have seen that major elements within the scientific community have become convinced that greenhouse gases do cause global warming and further, that something needs to be done about it *right now*. We have also seen that they considered a number of mitigation options and some of the most prestigious scientists and scientific organizations concluded (at taxpayer expense) that chemtrails would be the most cost-effective method. They also recognized that there would be environmental objections to and regulatory problems with this method.

I believe we can make a case that instead of them saying "oh well, we can't do it so let's try something else" they actually asked themselves "how can we hide this?" Yes, that leads us to a conspiracy theory or two, doesn't it? But, of course, the question remains, where is the proof? Unfortunately, if this is a vast international conspiracy, the perpetrators are so far extremely successful, as the only proof I have found after nearly ten years of searching has been the anecdotal and circumstantial.

ALASKA FLIGHT 261

Some of the patented Welsbach refractory agents have extreme abrasion characteristics, particularly aluminum oxide and silicon carbide. These materials have an extremely high hardness factor and are second only to diamond in abrasiveness. What effect might these 1-micron and sub-micron dusts have on aircraft traveling through the "grit-plume"?

The greases used on the horizontal and vertical stabilizers, ailerons, flaps, landing gear trucks and other surfaces inside the working flight components of aircraft could capture these particles, turning the grease into a kind of liquid sandpaper!

On 31 January 2000 Alaska Airlines Flight 261 crashed, killing all 88 passengers and crew on board. The crash site was off the California coast north of Los Angeles near Port Hueneme. The plane was en route from Puerto Vallarta, Mexico, to Seattle with a stop in San Francisco. Before the twin-engine MD-83 crashed, the pilot and co-pilot reported a problem with the plane's jackscrew, the mechanism that lifts or lowers the plane's nose in flight. They wrestled with the problem for more than 30 minutes before losing control. Around 4:19 PM Pacific Standard Time the plane rolled over and despite the valiant efforts of the pilots to right the craft, it slammed into the Pacific Ocean at 215 knots about 1 minute and 20 seconds later.

The National Transportation Safety Board investigation into the crash showed that the threads on the jackscrew were stripped. Several lawsuits were filed in the aftermath of the crash by relatives of the victims, claiming that the airline, the FAA, and the aircraft's manufacturer, Boeing, were all guilty of willful misconduct because they knew about problems with other jackscrews but did nothing about the jackscrew assembly on the plane that crashed. Industry officials countered, denying any wrongdoing and insisting that repeated tests showed the wear, prior to the crash, was insufficient to require replacement of the jackscrew assembly.

Dr. R. Michael Castle holds a National Certification for Environmental Risk Assessment with 15 years of field practice in Environmental Risk Assessment, Investigation, Analyses and Remediation. In his article "The Methodic Demise of Natural Earth Systems," Dr. Castle wrote:

> A horizontal-stabilizer Jack-Screw continuously coated with these highly abrasive dusts from the Welsbach Refractory materials will cause a gradual milling of the jack-screw metals and cause complete failure, jamming the flight controls into an uncontrollable down or up attitude configuration. We believe that Alaska Airlines Flight 261 was a victim of this unforeseen circumstance. Alaska Flight 261 made daily passage through heavy grit plumes from ChemTrails operations associated with Welsbach Refractory Seeding operations, principally along the West Coast of the US, down throughout Dallas, Texas.

Dr. Castle claims that other aircraft, commercial and military, have also suffered these flight-component failures, all of which have mistakenly been attributed to substandard workmanship by aviation machinists. "These conclusions," he wrote, "could not have been farther from the Truth in these matters."

You may know the adage, "the road to Hell is paved with good intentions," a pithy reminder that even the best meaning actions can result in very negative unexpected consequences. Could Alaska Flight 261 have been a victim of an unforeseen side effect of Welsbach seeding? Or, worse, could the chemtrail conspirators have foreseen this outcome and done little or nothing to stop it, considering a few airliners going down as inevitable "collateral damage"? What price might they be willing for us to pay for their playing God with our atmosphere?

"Deep Shield"

Brian Holmes raises a variety of poultry (everyone in his extended family gets a Christmas turkey!) on a farm called The Holmestead in the Township of Tiny, in the Province of Ontario, Canada. His website (www.Holmestead.ca) has been up for ten years. On it he relates the bucolic delights of life in the country, including descriptions of his own private wetlands. He was drawn into the chemtrail debate in the spring of 2002. After that he added many photographs, most taken at the Holmestead plus comments, observations and additional resources on chemtrails. Although it has visitors from around the world his site is intended as his personal attempt to raise local awareness of the chemtrail issue.

One of the most remarkable items on his website is an interview with a purported chemtrail insider. The person interviewed insisted on remaining anonymous so Brian dubbed him "Deep Shield." He was identified as an employee of Lawrence Livermore Labs. Brian called him "Deep Shield" in reference to the "Deep Throat" of Watergate fame and because that person called the chemtrail spraying program *The Shield Project*, although he said that that was not its official name. Some time after this interview occurred "Deep Shield" was reported to have committed suicide. This purported

286

chemtrail insider described himself saying:

> My official capacity is in direct research of atmospheric issues in relation to pollutants. I also create models of potential long-term effects of green house gasses on the climate. Predict wind patterns, weather patterns, etc. I have spent a good many years working on the [Shield] project calculating the amount of material needed and creating models for dispersion patterns. I work [with] other members who know the chemicals used and their interactions with the atmosphere, pollution and water vapor. I am part of a team which itself is part of a larger team, which is part of still a larger team.

I cannot prove that any of the following is true. If nothing else it makes for some very interesting reading. Here are some other excerpts from the interview, used with Brian's kind permission. See Brian Holmes's website for the full text:

> 1. What purpose do polymer threads imbedded with biological material serve in this scenario?
> Polymers are part of the mixture and they do form in threads and in `tufts'. The idea is simple and comes to us from the spider. As you may know spider webbing is very light, some newborn spiders spin a `parachute' to catch the prevailing breeze to travel far from their place of birth. Spiders have been able to attain high altitudes and travel great distances for long periods of time. Most of the elements used in the spray are heavier than air, even in their powdered form they are heavier and will sink quickly. Mixing them with the polymers suspends the particles in the atmosphere high above the surface for longer periods of time, therefore in theory we do not need to spray as often or as much material. Since the suspended particles eventually do settle into the lowest part of the atmosphere and

are inhaled by all life forms on the surface there is an attempt to counter the growth of mold by adding to the mixture mold growth suppressants—some of which may be of biological material.

Mold comes in spores that travel on the winds; the polymers can attract mold spores through static charges created by the friction of the polymer threads and the atmosphere. Add a bit of warmth and moisture and mold begins to grow. The polymer is stored in a liquid form as two separate chemicals. When sprayed they combine behind the plane `spinning' long polymer chains (threads). Much tinkering has been done with the chemical matrix in past years. Many polymers (plastics) are non-biodegradable thus add to the problem of pollution. Various formula have been used, some which even use biological agents.

2. If this spraying is to mitigate global warming, why does so much of it take place at night?

There is a desired concentration being sought. One that is thick enough to stem the UV and the Infrared, while being thin enough to allow visible light through. A perpetual cloud cover would have disastrous effects on plant life; the food chain thus disrupted would soon collapse. The desired effect wanted is a thin cover that would theoretically create a daytime haze that allows plenty of sunlight while providing protection from UV radiation and also reflect enough infrared to maintain nominal temperatures.

The different temperatures between day and night causes massive volumes of air to rise during the night, the warm air trapped at the surface rises above the cooling air above. By strategically spraying in certain areas at night, we get the advantage of the rising air, which not only pushes the material higher, but also causes the material to disperse into a thin layer.

...

4. What is the connection between ELF, EMF, VLF

and Chemtrails spraying? Or is there one?

To understand the use of radio waves in the shield, one first understands how ozone is created. I cannot stress to you how dire the situation really is. The shield in place is only a partial solution; we must counter the depletion of the ozone—this means we must make ozone in the stratosphere. Ozone at ground levels does no good; indeed, ozone pollution at ground levels it what is used to determine the air quality. Ozone is triatomic oxygen, O_3. It is the most chemically active form of oxygen. It is formed in the ozone layer of the stratosphere by the action of solar ultraviolet light on oxygen. Although it is present in this layer only to an extent of about 10 parts per million, ozone is important because its formation prevents most ultraviolet and other high-energy radiation, which is harmful to life, from penetrating to the earth's surface. Ultraviolet light is absorbed when its strikes an ozone molecule; the molecule is split into atomic and diatomic oxygen. Later, in the presence of a catalyst, the atomic and diatomic oxygen reunite to form ozone.

Ozone is also formed when an electric discharge passes through air; for example, it is formed by lightning and by some electric motors and generators. Since UV radiation is the problem, we can not use UV to produce more stratospheric ozone. Another method must be found. The shield acts like one plate of the electrode, when tickled with certain radio waves; it produces an opposite charge to stratospheric layers producing low atmosphere to stratosphere lightening. Creating ozone where it is needed.

5. If this is being done for the reasons you say, then why are other chemicals being used, why are different sprays being used?

Correcting the ecological damage that mankind has done has NEVER BEEN DONE BEFORE. We are relatively new to this notion of terraforming on a real

289

scale. That is what we are doing, Terraforming. We are trying to recreate the ideal life-sustaining conditions on a dying planet. We are testing and trying different methods.

Several attempts to improve the application of Shielding material and getting the most out of each application are taking place all the time. The combined resources of the nations of earth are not enough to allow constant spraying. Though we have achieved a high level of technology, there is a great surface area that needs to be covered nearly daily. Large sections of ocean are all but ignored; the remaining land masses are more than what can be covered effectively. The Shield would work best if it was a single thin layer without interruption, however due to the movement of air, weather patterns and the sad fact that we do not have the means to place ample amounts of material at the same level at the same time we are getting a small fraction of the effectiveness from our applications.

6. Why is spraying found before storm fronts? Is it to cause drought?

Before a storm there is a front, the front clears the air before a storm, pushing particulate matter ahead of it, leaving a space relatively clear of particulate matter. UV radiation levels rise in these areas, sometimes to dangerous levels. The shield must be maintained. Since barium absorbs water as well as carbon dioxide, precipitation has been affected. Other kinds of sprays are in development and testing which may reduce the affects on precipitation. As I stated above, this is a new technology we are working with, it is still in its infancy and there are some problems with it.

...

8. What about the reports of sickness after spraying?

There are several causatives for this. Some people are more sensitive to metals, while others are sensitive to the polymer chemicals. People will get sick, and some will die. It is estimated that 2 billion worldwide will be affected to some degree by the spraying. Without spraying we have a 90% + chance of becoming extinct as a species with in the next 20 years.

9. What is the relationship between these spraying programs and One World Order?

Personally I am against the move for globalization, and yes, there is potential to use the Shield to speed up the process of globalization, there are several countries that are involved in this project: European Union Nations, USA and Russia are the largest contributors to the project, many of the allied nations and UN Members participate to one extent or another. The material (chemical spray as you may call it) comes from all of these nations.

To insure that the chemicals are not tampered with, they are mixed and sprayed over random nations. This means that chemicals produced in the USA has a good chance of being sprayed over Russia, England and the USA. This random spray of material means that no nation would be certain that their chemicals will be sprayed over a nation which they have issues with. Russian planes may be seen in USA skies, but so too will USA planes be seen in Russian skies. The canisters used are sealed in a third nation that has no idea where its canister is going. Participating nations have their observers at every station where canister loading is done. All of this to insure that the shield is not used as a weapon. To further insure that the shield is not used as a weapon, non participant nations are sprayed by participants who must spray in order to get enough material to maintain their nations shield. It is understood that not spraying is as much a military offense as shooting at them.

Without the shield, UV poisoning would cause great death. The threat is a common one, which has brought nations together in defense. The natural outcome of having a common enemy is to strengthen international ties—a step toward globalization.

...

11. Why all the secrecy?

Due to the severity of the situation it is mandatory to maintain public calm for as long as possible. The Earth is dying. Humanity is on the road to extinction—without the Shield mankind will die off with in 20 to 50 years. Most people alive today could live to see this extinction take place. This means that an announcement of the situation we face boils down to telling every man, woman and child on earth that they have no future, they are going to be killed. People would panic. There would be economic collapse, the production and movement of goods would collapse. Millions would die in all cities on earth, riots and violence would reduce civilian centers to rubble within days. Half of the population in dense metropolitan areas would try to leave the cities seeking 'safety' in the rural areas thinking that they would be safe. Those left behind in the cities would be at war with their neighbors, fighting for the remaining supplies. We would be telling the world that the world is coming to an end, and even with the Shield the chances of survival are small.

UV Summer and Global Warming are the immediate problems we face, there are far greater problems that are raising their ugly heads and will present new problems which in some cases have no viable solutions at this time. Ecologies are collapsing. The extinction rate of species is climbing. The amount of chemical pollutants in the water and soil are fast approaching and in many places has surpassed the earth's ability to heal itself. Crop failure is on the rise, even in the USA the returns on crops are smaller than they were 10 years ago. Even with the advances in genetically altered food crops, we are falling behind in our ability to produce enough to go around. Throughout the 20th century chemical fertilizers and pesticides were used to insure the best yields. Unfortunately many of these have contaminated ground water, killed beneficial insects along with the undesirable insects. These chemicals have gotten into the food chain and are affected other species besides mankind. It is only a matter of years before famine spreads like a cancer throughout the world.

Clean fresh water is in short supply, in many places well water is non-potable, containing the run off of pesticides, herbicides and fertilizers that have been used on crops and lawns. The water treatment facilities we have are unable to scrub out all of the toxins we have placed in the soil and water supply. Many of the toxins we find build up over time in the body, a long slow poisoning which has been making its presence felt in many areas of the world in the form of cancers, leukemia, sterility, birth defects, learning disorders, immune deficiency problems, etc. These are on the rise, any good researcher can find the records. For decades there was public outcry for the end to pollution. For every small step we made to clean up our production, millions where born who added to the problem. Yes, pollution is down per individual, however there are a couple more billion individuals producing pollution, thus the real numbers have an increase in over all pollution produced. Name a city that does not have problems with smog. You would be hard pressed to find one. Though smog controls on automobiles is higher than ever before, the number of autos on the road has increased thus the amount of smog producing pollutants is higher than ever before. All the clean air acts passed to curb individual factory and auto emissions did not address the production of more factories and more autos. Here an uneasy

293

compromise was made between the need to maintain the economy against the need to maintain the ecology. The ecology lost since it was estimated to be a problem decades from now. The economy was a problem that would have dire effects today.

All of these factors combined have produced a scenario that in short boils down to the end of the world in 50 to 75 years. Even if we were to stop all emissions of pollution today, the inertia of past decades is enough to carry us over the brink in 100 years. However we cannot stop the production of pollution, to do so would mean shutting down every factory, every auto, every train, truck, ship and every household on the planet. Electricity is used to heat many homes in the Western World. The production of electricity produces fewer pollutants than heating all homes with wood or coal. Cutting our power generation abilities down to hydroelectric and fission reactors would leave a good chunk of the world in the dark. It is an impossible situation, our civilization is geared to the use of energy, take away our energy and civilization will collapse.

12. When will spraying stop?

There are several factors governing this:

A. Should the Ozone layer repair itself or our active attempts at repair reduces the amount of ground level UV to acceptable levels, spraying will stop. Present calculations place this between 2018 and 2024.

B. Should another method be found which is more effective, less costly or presents us with long-term solutions the Shield project would be replaced.

C. When the other problems become too big to make the maintenance of the shield worth the effort. The estimated date for this is 2025 to 2050.

13. Since Global Warming and UV summer are the problem, why is the Government backing down on its pollution controls?

Because they are ineffectual and will cause more

economic problems than they would solve ecological problems. We surpassed the threshold of Earth's ability to absorb pollutants in the 1970's. Since that time the earth's population has nearly doubled. Emerging Industrial nations have come into being, more pollutants are produced now than back then, even with the stringent controls in place. The world is heading for economic depression, more emission controls would add to the economic problems. This translates into our being unable to do anything to start solving the problems.

Unfortunately our technologies require a strong economy to advance. We need that advancement, we need the trillions of dollars spent on research that a strong economy causes. Each corporation that produces a product has a product development program in place. Many of the past products invented came by accident through other unrelated products. There is a corporate drive to find methods to clean up the ecology, to reduce emissions, etc. These goals have been in place for decades, many of the large corporations are in the know when it comes to the ecological problems we face thus they are spending a great deal of money and time on finding solutions to the problems we face. Take away the economy and their research stops.

...

24. Is all the spraying done using the "tank kits" described earlier or are the KC-135R and KC-10 types filled to the brim? Such aircraft have a load capacity of 200,000 pounds or more for refueling missions.

No. Several types of craft are used. Commercial jet airliners are used and they are not diverted from their flight paths to do so. How the canisters and the spraying is done on this kind of craft is unknown to me exactly. I do have my suspicions. I know best that which is my field; this is not to say that we do not talk around the water tank. So I know more than just my area and am able to think the matter through to its logical end. I do know that even all the commercial jetliners in use are not enough to insure complete coverage all of the time. My computer models require knowing how much material needs to be sprayed. Certain conditions cause wide areas to suddenly (over hours) open up in the Shield. Then and only then is mass spraying done - and would be done with the most logical craft, a tanker. Why not spray more from individual jetliners? That is one of the problems. Jetliners do not carry much material (100 to 500 gallons) because the material has to be spread out thinly. Look at the kinds of material being used, aluminum, barium, titanium, etc. Most are highly reflective; in some instances the material is an absorber of gasses. In the case of reflection the desire is to reflect X amount of heat and X amount of UV while still maintaining acceptable (nominal) levels of UV and heat reaching the planet's surface. Life requires a certain amount of both UV and heat too much will kill - so will too little. The apparent amount looks like a lot more than what is actually being sprayed per volume of air it is covering. Most of the whitening of the sky is not the material per se, but the collection of water vapor, which forms into suspended ice crystals. The introduction of the material causes the water vapor to collect like rain collects on individual particles of dust. Too much material would cause a "mud fall" of sorts where the naturally occurring water vapor would precipitate carrying the material with it. Spraying is done in such away as to "layer" the material through a volume that will allow an acceptable level of UV and heat through along with all the other wavelengths of light. Photosynthesis is the foundation of life on our planet. Only when all the material is removed in a local area does it require a massive spray, this is usually in the front of a weather system, or after a heavy period of precipitation. Then a tanker is flown, fully loaded.

Here let me inject another piece of "evidence." This has been circulating on the Internet since April of 2005. It may be a hoax. It was simply signed "A Concerned Citizen."

For reasons you will understand as you read this I can not divulge my identity. I am an aircraft mechanic for a major airline. I work at one of our maintenance bases located at a large airport. I have discovered some information that I think you will find important.

First, I should tell you something about the "pecking order" among mechanics. It is important to my story and to the cause to which you have dedicated yourself. Mechanics want to work on three things. The avionics, the engines, or the flight controls. The mechanics that work on these systems are considered at the top of the "pecking order". Next come the mechanics that work on the hydraulics and air conditioning systems. Then come the ones who work on the galley and other non-essential systems. But at the very bottom of the list are the mechanics that work on the waste disposal systems. No mechanic wants to work on the pumps, tanks, and pipes that are used to store the waste from the lavatories. But at every airport where I have worked there are always 2 or 3 mechanics that volunteer to work on the lavatory systems. The other mechanics are happy to let them do it. Because of this you will have only 2 or 3 mechanics that work on these systems at any one airport. No one pays much attention to these guys and no mechanic socializes with another mechanic who only works on the waste systems. Fact is, I had never even thought much about this situation until last month.

Like most airlines we have reciprocal agreements with the other airlines that fly into this airport. If they have a problem with a plane one of our mechanics will take care of it. Likewise, if one of our planes has a problem at an airport where the other airline has a maintenance base, they will fix our plane.

One day last month I was called out from our base to work on a plane for another airline. When I got the call the dispatcher did not know what the problem was. When I got to the plane I found out that the problem was in waste disposal system. There was nothing for me to do but to crawl in and fix the problem.

When I got into the bay I realized that something was not right. There were more tanks, pumps, and pipes then should have been there. At first I assumed that the waste disposal system had been changed. It had been about 10 years since I had worked on this particular model of aircraft. As I tried to find the problem I quickly realized the extra piping and tanks were not connected to the waste disposal system, at all.

I had just discovered this when another mechanic from my company showed up. It was one of the mechanics who usually works on this particular type of plane, and I happily turned the job over to him. As I was leaving I asked him about the extra equipment. He told me to "worry about my end of the plane and let him worry about his end!"

The next day I was on the company computer to look up a wiring schematic. While I was there I decided to look up the extra equipment I had found. To my amazement the manuals did not show any of the extra equipment I had seen with my own eyes the day before. I even tied in to the manufacturer files and still found nothing. Now I was really determined to find out what that equipment did.

The next week we had three of our planes in our main hanger for periodic inspection. There are mechanics crawling all over a plane during these inspections. I had just finished my shift and I decided to have a look at the waste system on one of our planes. With all the mechanics around I figured that no one would notice an extra one on the plane.

Sure enough, the plane I choose had the extra equipment! I began to trace the system of pipes, pumps, and tanks. I found what appeared to be the control unit for the system. It was a standard looking avionics control box but it had no markings of any kind. I could trace the control wires from the box to the pumps and valves but there were no control circuits coming into the unit. The only wires coming into the unit was a power connection to the aircraft's main power bus. The system had 1 large tank and 2 smaller tanks. It was hard to tell in the cramped compartment, but it looked like the large tank could hold about 50 gallons. The tanks were connected to a fill and drain valve that passed through the fuselage just behind the drain valve for the waste system.

When I had a chance to look for this connection under the plane I found it cunningly hidden behind a panel under the panel used to access the waste drain. I began to trace the piping from the pumps. These pipes lead to a network of small pipes that ended in the trailing edges of the wings and horizontal stabilizers. If you look closely at the wings of a large airplane you will see a set of wires, about the size of your finger, extending from the trailing edge of the wing surfaces. These are the static discharge wicks. They are used to dissipate the static electric charge that builds up on a plane in flight. I discovered that the pipes from this mystery system lead to every 1 out of 3 of these static discharge wicks. These wicks had been "hollowed out" to allow whatever flows through these pipes to be discharged through the fake wicks.

It was while I was on the wing that one of the managers spotted me. He ordered me out of the hanger telling me that my shift was over and I had not been authorized any overtime.

The next couple of days were very busy and I had no time to continue my investigation. Late one afternoon, two days after my discovery, I was called to replace an engine temperature sensor on a plane due to take off in two hours. I finished the job and turned in the paperwork. About 30 minutes later I was paged to see the General Manager. When I went in his office I found that our union rep and two others who I did not know were waiting on me. He told me that a serious problem had been discovered. He said that I was being written up and suspended for turning in false paperwork.

He handed me a disciplinary form stating that I had turned in false paperwork on the engine temperature sensor I had installed a few hours before. I was floored and began to protest. I told them that this was ridiculous and that I had done this work. The union rep spoke up at this point and recommended that we take a look at the plane to see if we could straighten it all out. I then asked who the other two men were. The GM told me that they were airline safety inspectors but would not give me their names.

We proceeded to the plane, which should have been in the air but was parked on our maintenance ramp. We opened the engine cowling and the union rep pulled the sensor. He checked the serial number and told everyone that it was the old instrument. We then went to the parts bay and went back into the racks. The union rep checked my report and pulled from the rack a sealed box. He opened the box and pulled out the engine temperature sensor with the serial number of the one I had installed. I was told that I was suspended for a week without pay and to leave immediately.

I sat at home the first day of my suspension

wondering what the hell had happened to me. That evening I received a phone call. The voice told me "Now you know what happens to mechanics who poke around in things they shouldn't. The next time you start working on systems that are no concern of yours you will lose your job! As it is, I'm feeling generous, I believe that you'll be able to go back to work soon." CLICK. Again, I had to pick myself from off the floor. As my mind raced, it was at this moment that I made the connection that what had happened to me must have been directly connected to my tracing the "mysterious" piping.

The next morning the General Manager called me. He said that due to my past excellent employment record that the suspension had been reduced to one day and that I should report back to work immediately. The only thing I could think of was "what are they trying to hide" and "who are 'THEY'"!

That day at work went by as if nothing had happened. None of the other mechanics mentioned the suspension and my union rep told me not to talk about it. That night I logged onto the Internet to try to find some answers. I don't remember now how I got there but I came across a site that talked about chemically-laced contrails. That's when it all came together. But the next morning at work I found a note inside my locked locker. It said, "Curiosity killed the cat. Don't be looking at Internet sites that are no concern of yours." Well that's it. Now I know 'THEY' are watching me.

While I don't know what THEY are spraying, I can tell you how they are doing it. I figure they are using the "honey trucks". These are the trucks that empty the waste from the lavatory waste tanks. The airports usually contract out this job and nobody goes near these trucks. Who wants to stand next a truck full of sh--. While these guys are emptying the waste tanks, it makes sense that they could easily be filling the tanks of the spray system. They know the planes flight path so they probably program the control unit to start spraying some amount of time after the plane reaches a certain altitude. The spray nozzles in the fake static wicks are so small that no one in the plane would see a thing. God help us all.

Now back to the Q & A with "Deep Shield" and a question directly relating to the letter quoted above.

25. Is there any truth in the story that some of the spraying is done by jetliners with modifications in the "honey" or waste compartment?

The technology used for spraying is rather simple. It requires at least two tanks under pressure, each carries half of the mixture which is sprayed at the same time forming a complete compound which is designed to be lightweight (so as to be suspended for longer periods of time). There have been attempts to incorporate the materials in jet fuel, however the material binds with unburned jet fuel, water vapor, etc and does not have the added buoyancy of the polymer threads. The end result is a spray that is less than half as effective and is more dangerous since it can lead to sulfates, acids and other mixtures, which are more lethal than the spray. It is very possible that the "honey" compartment is used. The amount of material needed is small compared to the payload of any given commercial airliner. However, there is a good deal of fuel tank that is not used. Airliners only fuel their craft for the journey ahead of them; they rarely top off the tank. This has become public knowledge in light of 9-11. It was this small fact that caused the terrorists to pick pan-continental flights so they would have a plane fully loaded. The majority of flights are short range and do not require the full capacity of an airliners fuel tanks. Any adaptations needed could easily be done during routine maintenance, and could be easily explained away as being a modification for safety and-or pollution controls. This last is my own theory. We can assume that any means possible to deliver the material is tried. Independent nations may favor one way of doing so over another.

...

27. Could you summarize the root causes of the initial destruction of the atmosphere that requires this "repair" work? Did it perhaps result in part from fluorides released/produced by the nuclear weapons programs?

In a word—Industry. Most fail to understand that the products we use, wear and live with are made in a manner that dumps CFC's and green houses gasses into the atmosphere. There is no one single causative in this issue. It goes way back to the Industrial Revolution and the use of coal to power steam engines. Since that time we have consumed greater and greater energy resources, dumping the waste where ever we wanted. Up until very recently refrigeration was a big contributor, imagine all those hundreds of millions of households that owned and operated freon cooled refrigerators from 1940 to 1970. Not just one refrigerator per household, but over the course of time often multiple freon units. This doesn't include the various air conditioner systems or industrial refrigeration systems. For a long period when the refrigerator or air conditioner unit was replaced, the old one was taken to the dump and thrown into the heap—the freon was free to escape and make its way up into the stratosphere to eat away at the ozone layer. You can add to that list. Think of all the cars that had air conditioners, think of all those hair spray cans with their propellant gasses—

303

the amount of those alone were enough to do great damage. Styrofoam is another industry and product that has contributed to the problem. In the scheme of things atomic energy has contributed little compared to the consumer goods that have been manufactured during the past century

So, are we humans the victims of our own success? Has the Industrial Revolution lead to the impending death of our planet? Or is all this just a lot of crazy talk and people buying into Internet hoaxes?

OTHER THEORIES

What about those other ideas of why we are being sprayed that I mentioned at the beginning of the last chapter? What about electromagnetic mind control schemes, or a delivery system for mass immunizations, or a delivery system for poisons to reduce global population to "sustainable" levels? Or even wilder, what about the idea that this geoengineering is really being done to change the Earth's atmosphere into something more like the home world of extraterrestrial invaders who will be moving in as soon as they can breathe what used to be our air?

Frankly, I don't want to go there. As you may have noticed, I have tried to avoid putting anything in this book that would not meet the basic requirements of a case at law. I have been trying to build my case against the military-industrial-academic complex on a base of scientifically established facts and credible, if circumstantial evidence. I have been very careful not to make any claims that I could not support with mainstream recognized citations. Yeah, I know, the whole "Deep Shield" stuff is hearsay and anecdotal, hardly "scientific" or "forensic" in nature—but it was cool, huh!?

I have read literally hundreds of emails and blog postings about these other claims as to who is producing chemtrails and why—and credible scientific evidence is just not there.

304

Keep in mind I am not saying that the ruling elite are not showering us with poison in an attempt to remove the useless eaters of the world, I just can't find any hard evidence to support the claim.

I can say that the scientific evidence is strongly against any theories that allege that this straying is in furtherance of any purposes to effect specific individuals or groups on the surface, such as through mass immunizations to defend Americans from air-bourn terrorist attacks or to eliminate undesirable minorities (or majorities) via ethno-specific biotoxins. That is because the higher in the atmosphere a substance is released the further it travels before reaching the ground, and the more dilute it becomes — thus making the targeting of those on the ground iffy at best. While there have been patents granted for aerosol vaccines those are for releases a few hundred feet above the surface, not in the stratosphere.

In April of 2001 a storm that kicked dust up into the jet stream in Mongolia dispersed dust from the Gobi Desert and industrial pollution from China across a quarter of the mainland United States and Canada. A whitish to brownish haze that reduced visibility less than 10 miles at times was seen from Calgary, Canada in the north, to Phoenix, Arizona in the south and as far east as the ski-runs of Aspen, Colorado. I was in Reno, Nevada at the time and clearly remember the near whiteout conditions downtown and the perplexed and amazed expressions on people's faces when they learned the dust came from 7,000 miles away.

Mount Pinatubo on the island of Luzon in The Philippines erupted catastrophically in June 1991 after 460 (+/-30) years of inactivity. More than 5 billion cubic meters of ash and pyroclastic debris were ejected, producing eruption columns 18 kilometers wide at the base and heights reaching up to 30 kilometers above the volcano's vent, sending this sunlight blocking material well into the stratosphere. For months the ejected volcanic materials remained suspended in the atmosphere where the winds dispersed them to envelope the earth. Some of this material circumnavigated the globe as many as three times before finally reaching the ground. This natural increase of the earth's albedo caused the world's temperature to fall by an average of 1 degree Celsius.

As mentioned above geoengineers calculated that artificial albedo enhancers would take three months to reach the surface if released into the lower troposphere, and a year or more if released into the stratosphere. How far would an aerosolized vaccine travel if released by a jetliner at flight altitude? A drug released from a United Airlines flight over Kansas would be "immunizing" people in the United Kingdom, Germany and Spain by the time it finally reached persons on the ground! And how concentrated would the material at the point of release have to be if the final dose were to be distributed over tens of thousands or millions of square miles? The same would be true of any sort of intentionally released poison.

No, whatever is going on up there is intended to do its thing up there and its coming to earth is only an inevitable by-product of that activity.

Of course that doesn't rule out a lot of the more wild sounding technological theories. Unfortunately, I am not technically versed enough to be able to evaluate most of these. Surely some of the materials being injected into the upper atmosphere are part of radio, radar and similar electromagnetic operations. I have no problem believing that much of what's going on has to do with advanced radar and other communications and surveillance systems. Which would be more of the military's use of natural systems to achieve DoD purposes – i.e. weaponizing the atmosphere. One could argue that keeping these secret would legitimately fall under National Security. Of course the question arises at what point does the right of the military to defend the "health" of the nation override the right of the citizens to breathe healthful air? Are they keeping these operations secret from our enemies — or from our lawyers?

Over the last decade I have been asked hundreds of times if there is a link between HAARP and chemtrails. I have posed that question to many of the top chemtrail researchers such as

Clifford Carnicom and Will Thomas — Clifford said "yes" and Will said "no"!

There are a few possible modes of use for HAARP that could include chemtrails (if they really exist), but I can't say with any certainty that HAARP is actually are being used for any of these. Which is to say these are dots that *could* be connected but there is no hard evidence that they *are* connected.

Roughly, one would be in using the energy of the HAARP signal pushing up under the bottom of the particle cloud to help keep it aloft. U.S. Patent 4253190 describes how a mirror made of "polyester resin" could be held aloft by the pressure exerted by electromagnetic radiation from a transmitter like HAARP. If they really are doing geoengineering to mitigate global warming then the artificial albedo material, which is heavier than air, could use assistance in be kept up as long as needed.

Another possibility is using the bottom of the particle cloud to bounce transmissions off of, as mentioned earlier. A NATO paper, "Modification of Tropospheric Propagation Conditions" published in May 1990 detailed how the atmosphere could be modified to absorb electromagnetic radiation by spraying polymers behind high-flying aircraft. DARPA's admission that they have used HAARP to achieve "control of electron density gradients and the refractive properties in selected regions of the ionosphere to create radio wave propagation channels" shows that they have an interest in creating artificial ionospheric bounce spots. Now, as to whether they are also using chemtrails to create such bounce spots lower in the atmosphere is anybody's guess. This may also be one form of over the horizon radar. If mind control were taking place this could be one of the modes of accomplishing it as hinted at in the writings of Dr. Gordon J. F. MacDonald. Do you *feel* mind controlled?

In the end you will have to do your own research — and believe what you want to believe — at least until whoever's doing it officially admits to what they've been up to and why.

CHAPTER EIGHT

CONCLUSION

"Never believe anything until it's officially denied."
~ **Margaret Atwood**

There are men in all ages who mean to govern well, but they mean to govern. They promise to be good masters ... but they mean to be masters.
~ **Daniel Webster**

I want to end this book with an interesting email, one that focuses on exactly what I had been planning to write about in this chapter. Perfectly timed, this came in at the end of September 2006, just as I was wrapping up this book. It is from a long-time friend of mine (who unfortunately wishes to remain anonymous). He has been researching HAARP and weather modification almost as long as I have, and we have been chatting about it via email for many years. He doesn't have any specific training in this area, just a brilliant mind (certified genius). Like all of us researching this he has had to play connect the dots with bits of scientific evidence, news reports and direct observation. Maybe you've come to similar conclusions:

> I'm sure you remember my contention that the reason they rushed HAARP up to full-power in late spring this year was so that they could use it to generate a thermal shield over the southeast to ward off hurricanes, in order to avoid another Katrina-type disaster just months before this November's elections? While I have no way of obtaining "hard proof", you can see that my theory is holding up well. The season

is almost over and there has not been 1 single hurricane make landfall on the continental U.S. this year, expect for the first which actually dropped from hurricane status before it hit land. Compare this unbelievable statistic of "none" to last year's record-breaking season. Compare the seasonal track maps for 2005 and 2006 and tell me if this year looks remotely "natural" compared to last year: http://www.nhc.noaa.gov/2005atlan.shtml map at the bottom of the page for 2005. Scroll down the page and you can see the tracks for each storm so far this year at: http://enwikipedia.org/wike/2006_Atlantic_hurricane_season.

Note the pattern of behavior from earliest to latest as the shield built up strength. Basically all of the tropical storms this year except the first (when the shield was still building up) have veered north and east before getting to the U.S. coast. One (called "Chris") was stopped dead in its tracks just east of Cuba when it slammed head-on into the shield and just dissipated! That *never* happened last year (or any other year that I can recall). Each and every one of them, bumped into the shield (officially an "abnormally persistent zone of high pressure hovering over the North American continent") and been shunted off harmlessly into the North Atlantic this year. When last year most of them just barreled straight on into the Gulf, storm after storm. I don't see how anyone can explain this total shift in the behavior of the hurricanes in the mere span of a single year as "natural". Particularly when this year's season is so very different for all previous seasons, not just 2005.

Also, last year was a record-breaking year for the number of super strong hurricanes—cat 4 and 5, with the most cat 5 ever. But this year there are fewer hurricanes and they all much weaker. Why? Because the majority of tropical storms only become hurricanes when they enter the warm waters of the Caribbean and the Gulf—that's where they really grow huge and pick up power. If you block them from entering the Caribbean and the Gulf, you keep the size and strength way down—and you prevent many "tropical storms" from ever developing into true "hurricanes" in the first place.

Circumstantial "evidence" for my speculation:

1) They rushed HAARP to full power ahead of schedule, just in time to have it finished by the start of this year's hurricane season. Why the sudden rush to get to full power ahead of schedule?

2) There's the total dichotomy between the behavior of the hurricanes this season compared to last year (and the all the years before).

3) There was historically record-shattering heat across the nation this summer. Think about it... if you heat up the atmosphere over the lower part of the continent, you may well create an artificially-persistent high pressure zone that bounces hurricanes into the North Atlantic, but you can't avoid over-heating of the air beneath the shield at the same time.

4) About 3 weeks ago there was a 6.0 earthquake centered in the Gulf. Luckily, its location and strength wasn't quite big enough to generate a tsunami, but consider that the Gulf is incredibly stable geologically speaking. In the time they have been recording earthquakes on this continent, there have only be 30 earthquakes in the Gulf... and most all have been very small... 2 or 3 on the scale. This 6.0 was by far the biggest Gulf quake in 33 years—and it happened in conjunction with 1), 2) and 3) above. The "coincidences" are mounting up into a pretty good endorsement of my speculation, I think.

Links on the quake: http://earthquake.usgs.gov/eqcenter/receteqsww/Quakes/usslav.php#maps

and

http://news.nationalgeographic.com/

311

news/2006/09/060911-earthquake.html.

So... that's a fair amount of circumstantial evidence to support my theory.

While the pattern of this 2006 hurricane season in the Atlantic is completely different from last year (and all other previous years), it is important to note that the hurricane pattern and season in the *Pacific* is meeting the predictions that it would be another bad year, just as 2005 was, and the tracking patterns of the hurricanes in the Pacific are following their usual historic routes. So, we have a "normal" hurricane season/pattern in the Pacific but a completely "abnormal" (and highly beneficial to the U.S.) season/pattern in the Atlantic—again anyone trying to make sense of these two dichotomies has to ask Why and How they could happen?

You have just finished reading over 300 pages of similar circumstantial evidence. Let me remind you of a few of the dots I consider important. Remember Dr. Ross N. Hoffman from Atmospheric and Environmental Research (AER)? He wrote:

> If it is true, as our results suggest, that small changes in the temperature in and around a hurricane can shift its path in a predictable direction or slow its winds, the question becomes, How can such perturbations be achieved. ... It might be possible ... to heat the air around a hurricane and thus adjust the temperature over time. ... Our team plans to conduct experiments in which we will calculate the precise pattern and strength of atmospheric heating needed to moderate hurricane intensity or alter its track.

Is it possible that the AER team (or anyone else, for that matter) has in fact completed this work and now knows exactly how to shift a hurricane in its path and/or how to slow

312

its winds? Was HAARP, or some other as yet unidentified facility, responsible for saving the coast while scorching the nation?

Let me also remind you of Joe Gelt's comments in his 1992 article for *Arroyo* magazine, "Weather Modification: A Water Resource Strategy to be Researched, Tested Before Tried," in which he pointed out that:

> …weather modification … goes against the grain of a certain ecological ethic. It represents an interference with a natural process, with results possibly difficult to predict and control. Man as geologic force built dams and controlled the course of powerful rivers, upsetting along the way ecological balances and causing environmental harm. What then might man as an atmospheric force accomplish?

> Oft-lamented is the fact that everybody talks about the weather but no one does anything about it. To do something about the weather however is to raise various complex legal and public policy questions. For example: Who is liable for damages from floods or other weather events resulting from weather modification? How are the rights of those who want rain to be reconciled with the rights of those who prefer sunshine…

How are we to balance the benefit of a coastline not ravaged by hurricanes against a heartland withered by heat? To whom do the farmers send their bill? To whom do the coastal property owners, and their beleaguered insurance agents, send their thanks? And what about the environment? Coastal waterways need the storm surge of a hurricane to refresh them. If we have many years like this one those coastal sloughs will become stagnant stinking swamps. Are the perpetrators of this ecological crime keeping their name out of the papers to avoid legal action, as usual? Are we really

313

ready to let arrogant academics and/or a power-tripping military play God, or err, Goddess?

What about that earthquake in the Gulf my friend mentioned? As I previously quoted from *Planet Earth, The Latest Weapon of War*, Dr. Rosalie Bertell wrote:

> Ionospheric heaters such as HAARP create extremely low frequency (ELF) waves which are reflected back to the Earth by the ionosphere. These rays can be directed though the Earth … ELF waves can be used to convey mechanical effects, vibrations, at great distances … it has the capability to cause disturbance of volcanoes and tectonic plates, which in turn have an effect on the weather.

Weather + earthquakes:

It is quite possible that electromagnetic (radio) waves can be used to heat more than just the ionosphere, and in doing so may create a variety of effects from above the atmosphere all the way down to deep underground. Or, as U.S. Secretary of Defense William S. Cohen put it in 1997 "they can alter the climate, set off earthquakes, volcanoes remotely through the use of electromagnetic waves."

A question of keen interest then becomes who are the "they" Cohen is talking about? We can see a parable to the answer to that question in the story of Jesus casting the demons out of the possessed man and into a herd of pigs. In the King James Version of Mark 5:9 we read: "And [Jesus] asked him, What is thy name? And he answered, saying, My name is Legion: for we are many." Or, as Tears For Fears sang it "Everybody Wants To Rule The World." There are indeed legions of players in this game. Some do it for profit, some out of altruism, some to further their careers, some meddle with Mother Nature just 'cuz they can, and others do it for domination.

Keith Harmon Snow concluded his massive report "Out of the Blue: Black Programs, Space Drones & the Unveiling of U.S. Military Offensives in Weather as a Weapon" with a list of conclusions that could be drawn from his research. Let me list here the ones that I agree with, or at least are supported by my research:

The general public remains confused by climate skeptics.

The scientific community is mostly engaged in a narrow debate about climate change.

The spectrums of problems of climatic mayhem are greatly unappreciated.

Where these problems are appreciated, proponents argue narrowly about fossil fuels and climate protocols that, conveniently, distract and deflect attention from greater issues of secrecy, military dominance and environmental chaos.

Military and "civilian" ENMOD capabilities are already being tested, and quite likely have already been deployed to affect massive human loss of life and environmental instability.

The U.S. government position vacillates between admissions that limited development of ENMOD technologies has occurred — always in the private sector — and that ENMOD technologies do not exist at all.

Scientists, soldiers and government officials have lied outright, and many continue to intentionally obfuscate and misinform on climate issues and weather warfare.

The duplicitous and self-interested positions of these skeptics and liars attain the greatest currency through the corporate mass and entertainment media.

There is a trillion dollar industry behind the moneyed interests, and the propaganda, of fossil fuels, weather warfare, military and climate issues.

ENMOD and weather weaponry relies on widespread environmental instability to provide a threshold of "background" chaos to shield its covert ENMOD operations.

Thus we are unable to determine the extent to which changes have occurred, through ENMOD manipulations, or the extent to which they are occurring, or in future will occur.

The United States of America has violated the 1977 Convention on the Prohibition of Military or Other Hostile Use of Environmental Modification Techniques, (ENMOD Treaty), of which it is a long-standing signatory member.

The United Nations has demonstrated its lack of attention and investigation into climate issues and the violations of international treaties ...

Rich and poor countries alike will increasingly suffer as accelerated processes of environmental change are aggravated by unforeseen feedback mechanisms.

The radical shift to an alternate state or states of climate, most probably undesirable and unmanageable, has become an increasingly likely—if not certain— event, and it is increasingly likely that such an event will occur sooner rather than later.

Snow's point is simple, if broad. Those engaged in military EnMod are supported by a vast network of the rich and powerful whose objectives are served through the use of EnMod and their corporately owned media's careful failure to inform us about it. Further, the whole greenhouse gases academic argument is little more than a smokescreen behind which massive environmental damage is being done in an academic parallel to the legend of Nero fiddling while Rome burned— scientists argue while environmental damage grows. Further, the damage from global warming is actually being intentionally increased because "ENMOD and weather weaponry relies on widespread environmental instability to provide a threshold of "background" chaos to shield its covert ENMOD operations." The real chemtrails con could be that while pretending to save the world they are actually destroying it.

Snow believes that the major forces at work here are little more than a refinement of the policies of the Robber Baron monopolists of the 19th century—control the resources and reduce third world populations to useful levels (just enough to do the needed work, and no more).

I think there's more to it than this. The common thread through all my writing is that I write about those I call "control freaks." You take any 100 people and put them into a room. A handful will instinctively "rise" to be the leaders—some people feel compelled to rule. Another handful of the people in that room will just as instinctively want nothing to do with these self-proclaimed rulers (I'm afraid I'm one of them). The rest of the folks in the room will go along to get along, believing that someone has to rule and they don't want it to be themselves. Well, now that the human family has grown to over six billion of us that means there are several hundred million "control freaks" out there needing someone—or some thing, like the weather—to control.

With the control freaks it really doesn't matter why, *how* to achieve control is all that it is important. There are therefore many objectives of those behind this, as the object is to attain and maintain a position of power—and there are a near infinity of such positions, from being the "boss" of the mail room to being the president of the World Bank. We have many levels of competition in this world from corporations to governments to religions. Who doesn't want to get "a handle" on the competition?

As previously mentioned, Dr. MacDonald wrote that in the future:

> ...technology will make available to the leaders of the major nations a variety of techniques for conducting secret warfare... techniques of weather modification could be employed to produce prolonged periods of drought or storm, thereby weakening a nation's capacity and forcing it to accept the demands of the competitor.

317

In *Between Two Ages* Brzezinski openly discussed conducting war covertly by causing droughts, storms, volcanic eruptions and floods. We can see that between nations weather warfare would be a very useful handle.

But national leaders may not be on the highest level of the pyramid of power. The true capstone would be to rule the world—but to do that you need a world government with sufficient power to make the people of the world obey. I have long believed that Agenda 21 that came out of the Rio Earth Summit in 1992, and now the Earth Charter, were but part of a grand scheme to create such a world government—and global warming and its seemingly attendant chaotic weather plays into the hands of those who would use it as a pretext to build that capstone, through such international agreements as the Kyoto Protocols.

One of the great selling points of world government is the idea that since wars are fought between nations, if eliminate nations then you eliminate wars. Now here's an evil choice for you—which would you prefer: endless ever-more-destructive wars between half-wits running superpowers, or a global tyrant with powers that would make Hitler green with envy?

Can we create a Civil Society that can make a world without nations and without an ultimate Master? Can we save our world from over-population and under-management without resorting to creating a Green Gestapo and mass graves? Can we ever learn the truth about what the military-industrial-academic complex is up to before it is too late?

I have learned that the absolute worst possible thing never happens—so I always expect it. Why? Because when it doesn't happen I am pleasantly surprised, and being pleasantly surprised is so much better than being disappointed! So I am not going to expect any of the answers to the questions in the previous paragraph to be "yes"—feel free to surprise me!

Top: The HAARP antenna array in Alaska. Bottom: How the antenna array is laid out on the ground.

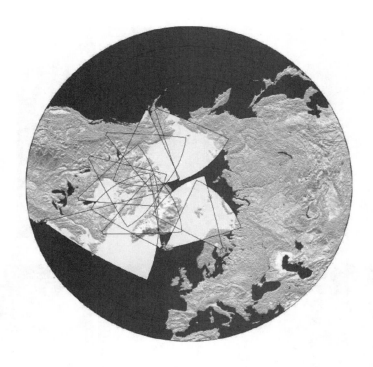

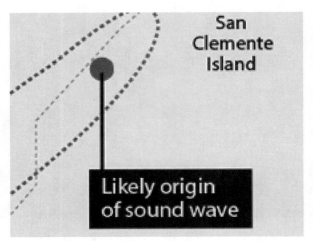

Top: The Super DARN map showing how the HAARP antenna array could cover much of the northern polar area. Bottom: Diagram of the unusual sound wave discovered near San Diego, California.

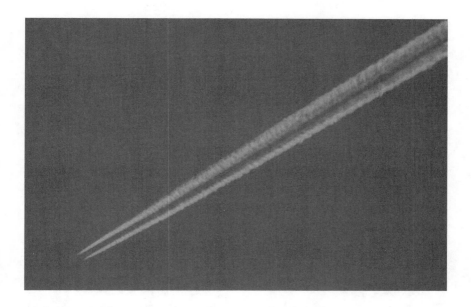

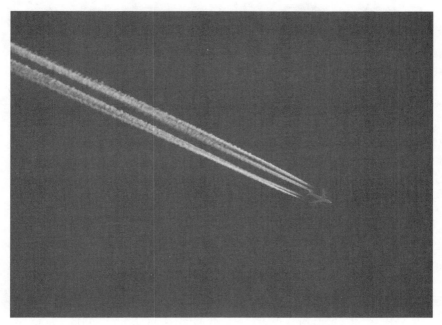

Top: Chemtrail or Contrail?—That is the question. Bottom: Possible chemtrails over western Minnesota, March 2003. Photos by Jaye Beldo.

Chemtrails 21 May 2002 Georgian Bay, Ontario - © 2002 G. Brian Holmes

Top: Possible chemtrails over western Minnesota, March 2003. Photo by Jaye Beldo. Bottom: Possible chemtrails over Georgian Bay, Ontario. Photo by Brian Holmes.

Top: Weather warfare could involve artifical lightning strikes. Bottom: Melbourne, Australia photo of unusual clouds, possibly being modified by a scalar-wave weapon or other "weather warfare" instrument.

The entire Earth becomes the battleground in future scenarios of weather warfare. With the push of a button catastrophes can be created.

Top: During the 2005 Florida hurricane *Wilma* this unusl "2" shape was captured by weather radar. Bottom: Artifical tornados and hurricanes could be created with weather modification equipment.

325

Weather warfare can also include space-based weapons to be used for weather modification.

ABOUT THE AUTHOR

Jerry E. Smith has been a writer, editor and activist since the 1960s. His bibliography of published works includes scores of non-fiction articles and reviews, more than a dozen ghost-written books, and two previous non-fiction books from Adventures Unlimited Press (AUP): *HAARP: The Ultimate Weapon of the Conspiracy* (1998) and *SECRETS OF THE HOLY LANCE: The Spear of Destiny In History & Legend* (2005). The latter was co-authored with George Piccard, author of *LIQUID CONSPIRACY: JFK, LSD, the CIA, Area 51, and UFOs* (AUP 1999). Since HAARP's North American release in 1998 it has been translated into Portuguese, in 2005, by Editora Aleph of São Paulo, Brazil, as *ARMAS ELETROMAGNÉTICAS: seria o projeto Haarp a próxima ameaça mundial?* and in 2001 into Polish by Amber Supermedia as *HAARP BRON OSTATECZNA*.

Jerry E. Smith is also a popular TV and radio talk show guest and lecturer. He has been on over 100 different radio shows (many times as a repeat guest) and on over a dozen TV shows, including *Encounters With The Unexplained* on the PAX TV Network, *Unscrewed* on the Tech TV Network and *Decoding The Past: Relics of The Passion* on The History Channel in the US and on *Conspiracies* on SkyOne Satellite in the UK. He has lectured across the US and as far away as The Bahamas, Brazil and Brisbane, Australia.

Jerry E. Smith began his career in writing and publishing in the little magazine field in 1966, writing and publishing his own amateur sci-fi fan magazine (called a fanzine or just "zine") in the Valley Science Fiction Associations' *Valley Amateur Press Alliance (ValAPA)* of Pomona, California. He was active in the zine scene throughout the late '60s and early '70s, appearing regularly in zines and APAs, like the Los

Angeles Science Fantasy Society's *APA-L*, and culminating with his founding of the Unicorn Society and its *Unicorn Amateur Press Alliance (UnAPA)* in Klamath Falls, Oregon, in 1974.

He was a close friend and literary partner of the late Jim Keith (author of *Mind Control, World Control*, and many others). Jim was Jerry's best friend from high school till his untimely death in 1999. Among their many projects together, they ran the regional newspaper *Skyline: Klamath Falls*; and co-hosted a radio show broadcasting from the campus of the Oregon Institute of Technology. Jerry helped on several of Jim's books, particularly *Black Helicopters Over America: Strikeforce for the New World Order* and *Secret And Suppressed: Banned Ideas & Hidden History*.

In 1990 Jerry and Jim founded the National UFO Museum (NUFOM) in Reno, Nevada. Jerry worked as the Executive Director, while Jim acted as the Chairman of the Board. In addition to his administrative duties of running the day-to-day operations of NUFOM, Mr. Smith also edited and wrote for that organization's quarterly journal, *Notes from the Hangar*. At the same time Jerry worked as an editor/graphic artist with Jim Keith's magazine *Dharma Combat: The Magazine of Spirituality, Reality and Other Conspiracies (DC)*, where Jerry served variously as Managing Editor and Art Director from its inception in 1988 till Jim's death. Jerry also wrote for DC under the penname of jarod o'danu.

Mr. Smith is divorced with one son. Religiously, he describes himself as "a cross between a non-practicing Taoist and a quantum physicist who can't do the math." On the other hand, he says his brand of politics is "straight Jeffersonian Libertarian with just a touch of anarchist leanings."

Please visit his website at: www.jerryesmith.com. Jerry is always happy to read and tries to reply to all email from his readers! You can reach Jerry at jerry@jerryesmith.com.

APPENDIX A

CONVENTION ON THE PROHIBITION OF MILITARY OR ANY OTHER HOSTILE USE OF ENVIRONMENTAL MODIFICATION TECHNIQUES

Adopted by the United Nations General Assembly on 10 December 1976. The Convention opened to signatures at Geneva on 18 May 1977 and entered into force 5 October 1978.

The States Parties to this Convention,

Guided by the interest of consolidating peace, and wishing to contribute to the cause of halting the arms race, and of bringing about general and complete disarmament under strict and effective international control, and of saving mankind from the danger of using new means of warfare,

Determined to continue negotiations with a view to achieving effective progress towards further measures in the field of disarmament,

Recognizing that scientific and technical advances may open new possibilities with respect to modification of the environment,

Recalling the Declaration of the United Nations Conference on the Human Environment adopted at Stockholm on 16 June 1972,

Realizing that the use of environmental modification techniques for peaceful purposes could improve the interrelationship of man and nature and contribute to the preservation and improvement of the environment for the benefit of present and future generations,

Recognizing, however, that military or any other hostile use of such techniques could have effects extremely harmful to human welfare,

Desiring to prohibit effectively military or any other hostile use of environmental modification techniques in order to eliminate the dangers to mankind from such use, and affirming their willingness to work towards the achievement of this objective,

Desiring also to contribute to the strengthening of trust among nations and to the further improvement of the international situation in accordance with the purposes and principles of the Charter of the United Nations,

Have agreed as follows:

Article I

1. Each State Party to this Convention undertakes not to engage in military or any other hostile use of environmental modification techniques having widespread, long-lasting or severe effects as the means of destruction, damage or injury to any other State Party.

2. Each State Party to this Convention undertakes not to assist, encourage or induce any State, group of States or international

330

organization to engage in activities contrary to the provisions of paragraph 1 of this article.

Article II

As used in Article I, the term "environmental modification techniques" refers to any technique for changing -- through the deliberate manipulation of natural processes -- the dynamics, composition or structure of the Earth, including its biota, lithosphere, hydrosphere and atmosphere, or of outer space.

Article III

1. The provisions of this Convention shall not hinder the use of environmental modification techniques for peaceful purposes and shall be without prejudice to the generally recognized principles and applicable rules of international law concerning such use.

2. The States Parties to this Convention undertake to facilitate, and have the right to participate in, the fullest possible exchange of scientific and technological information on the use of environmental modification techniques for peaceful purposes. States Parties in a position to do so shall contribute, alone or together with other States or international organizations, to international economic and scientific co-operation in the preservation, improvement, and peaceful utilization of the environment, with due consideration for the needs of the developing areas of the world.

Article IV

Each State Party to this Convention undertakes to take any measures it considers necessary in accordance with its constitutional processes to prohibit and prevent any activity in violation of the provisions of the Convention anywhere under its jurisdiction or control.

Article V

1. The States Parties to this Convention undertake to consult one another and to cooperate in solving any problems which may arise in relation to the objectives of, or in the application of the provisions of, the Convention. Consultation and cooperation pursuant to this article may also be undertaken through appropriate international procedures within the framework of the United Nations and in accordance with its Charter. These international procedures may include the services of appropriate international organizations, as well as of a Consultative Committee of Experts as provided for in paragraph 2 of this article.

2. For the purposes set forth in paragraph 1 of this article, the Depositary shall, within one month of the receipt of a request from any State Party to this Convention, convene a Consultative Committee of Experts. Any State Party may appoint an expert to the Committee whose functions and rules of procedure are set out in the annex, which constitutes an integral part of this Convention. The Committee shall transmit to the Depositary a summary of its findings of fact, incorporating all views and information presented to the Committee during its proceedings. The Depositary shall distribute the summary to all States Parties.

3. Any State Party to this Convention which has reason to believe that any other State Party is acting in breach of obligations deriving from the provisions of the Convention may lodge a complaint with the Security Council of the United Nations. Such a complaint should include all relevant information as well as all possible evidence supporting its validity.

4. Each State Party to this Convention undertakes to cooperate

in carrying out any investigation which the Security Council may initiate, in accordance with the provisions of the Charter of the United Nations, on the basis of the complaint received by the Council. The Security Council shall inform the States Parties of the results of the investigation.

5. Each State Party to this Convention undertakes to provide or support assistance, in accordance with the provisions of the Charter of the United Nations, to any State Party which so requests, if the Security Council decides that such Party has been harmed or is likely to be harmed as a result of violation of the Convention.

Article VI

1. Any State Party to this Convention may propose amendments to the Convention. The text of any proposed amendment shall be submitted to the Depositary who shall promptly circulate it to all States Parties.

2. An amendment shall enter into force for all States Parties to this Convention which have accepted it, upon the deposit with the Depositary of instruments of acceptance by a majority of States Parties. Thereafter it shall enter into force for any remaining State Party on the date of deposit of its instrument of acceptance.

Article VII

This Convention shall be of unlimited duration.

Article VIII

1. Five years after the entry into force of this Convention, a conference of the States Parties to the Convention shall be convened by the Depositary at Geneva, Switzerland. The conference shall review the operation of the Convention with

a view to ensuring that its purposes and provisions are being realized, and shall in particular examine the effectiveness of the provisions of paragraph 1 of Article I in eliminating the dangers of military or any other hostile use of environmental modification techniques.

2. At intervals of not less than five years thereafter, a majority of the States Parties to the Convention may obtain, by submitting a proposal to this effect to the Depositary, the convening of a conference with the same objectives.

3. If no conference has been convened pursuant to paragraph 2 of this article within ten years following the conclusion of a previous conference, the Depositary shall solicit the views of all States Parties to the Convention, concerning the convening of such a conference. If one third or ten of the States Parties, whichever number is less, respond affirmatively, the Depositary shall take immediate steps to convene the conference.

Article IX

1. This Convention shall be open to all States for signature. Any State which does not sign the Convention before its entry into force in accordance with paragraph 3 of this article may accede to it at any time.

2. This Convention shall be subject to ratification by signatory States. Instruments of ratification or accession shall be deposited with the Secretary-General of the United Nations.

3. This Convention shall enter into force upon the deposit of instruments of ratification by twenty Governments in accordance with paragraph 2 of this article.

4. For those States whose instruments of ratification or accession are deposited after the entry into force of this

Convention, it shall enter into force on the date of the deposit of their instruments of ratification or accession.

5. The Depositary shall promptly inform all signatory and acceding States of the date of each signature, the date of deposit of each instrument of ratification or accession and the date of the entry into force of this Convention and of any amendments thereto, as well as of the receipt of other notices.

6. This Convention shall be registered by the Depositary in accordance with Article 102 of the Charter of the United Nations.

Article X

This Convention, of which the English, Arabic, Chinese, French, Russian, and Spanish texts are equally authentic, shall be deposited with the Secretary-General of the United Nations, who shall send certified copies thereof to the Governments of the signatory and acceding States.

IN WITNESS WHEREOF, the undersigned, being duly authorized thereto by their respective governments, have signed this Convention, opened for signature at Geneva on the eighteenth day of May, one thousand nine hundred and seventy-seven.

DONE at Geneva on May 18, 1977.

ANNEX TO THE CONVENTION
CONSULTATIVE COMMITTEE OF EXPERTS

1. The Consultative Committee of Experts shall undertake to make appropriate findings of fact and provide expert views

relevant to any problem raised pursuant to paragraph 1 of Article V of this Convention by the State Party requesting the convening of the Committee.

2. The work of the Consultative Committee of Experts shall be organized in such a way as to permit it to perform the functions set forth in paragraph 1 of this annex. The Committee shall decide procedural questions relative to the organization of its work, where possible by consensus, but otherwise by a majority of those present and voting. There shall be no voting on matters of substance.

3. The Depositary or his representative shall serve as the Chairman of the Committee.

4. Each expert may be assisted at meetings by one or more advisers.

5. Each expert shall have the right, through the Chairman, to request from States, and from international organizations, such information and assistance as the expert considers desirable for the accomplishment of the Committee's work.

APPENDIX B

HAARP PATENTS

These were all originally assigned to APTI, Inc. and are now presumed to be the property of BAE Systems North America.

U.S. Patent 4686605:
Method And Apparatus For Altering A Region In The Earth's Atmosphere, Ionosphere, And/Or Magnetosphere
Inventors: Eastlund; Bernard J., Spring, TX
Assignees: APTI, Inc., Los Angeles, CA
Issued: Aug. 11, 1987
Filed: Jan. 10, 1985

U.S. Patent 5038664:
Method For Producing A Shell Of Relativistic Particles At An Altitude Above The Earth's Surface
Inventors: Eastlund; Bernard J., Spring, TX
Assignees: APTI, Inc., Washington, DC
Issued: Aug. 13, 1991
Filed: Jan. 10, 1985

U.S. Patent 4712155:
Method And Apparatus For Creating An Artificial Electron Cyclotron Heating Region Of Plasma
Inventors: Eastlund; Bernard J., Spring, TX
 Ramo; Simon, Beverly Hills, CA
Assignees: APTI, Inc., Los Angeles, CA
Issued: Dec. 8, 1987
Filed: Jan. 28, 1985

U.S. Patent 5068669:

Power Beaming System
Inventors: Koert; Peter, Washington, DC
 Cha; James T., Fairfax, VA
Assignees: APTI, Inc., Washington, DC
Issued: Nov. 26, 1991
Filed: Sep. 1, 1988

U.S. Patent 5218374:
Power Beaming System With Printer Circuit Radiating
Elements
 Having Resonating Cavities
Inventors: Koert; Peter, Washington, DC
 Cha; James T., Fairfax, VA
Assignees: APTI, Inc., Washington, DC
Issued: June 8, 1993
Filed: Oct. 10, 1989

U.S. Patent 5293176:
Folded Cross Grid Dipole Antenna Element
Inventors: Elliot; Paul G., Vienna, VA
Assignees: APTI, Inc., Washington, DC
Issued: Mar. 8, 1994
Filed: Nov. 18, 1991

U.S. Patent 5202689:
Lightweight Focusing Reflector For Space
Inventors: Bussard; Robert W., Manassas, VA
 Wallace; Thomas H., Gainesville, FL
Assignees: APTI, Inc., Washington, DC
Issued: Apr. 13, 1993
Filed: Aug. 23, 1991

U.S. Patent 5041834:
Artificial Ionospheric Mirror Composed Of A Plasma Layer
 Which Can Be Tilted
Inventors: Koert; Peter, Washington, DC
Assignees: APTI, Inc., Washington, DC

Issued: Aug. 20, 1991
Filed: May. 17, 1990

U.S. Patent 4999637:
Creation Of Artificial Ionization Clouds Above The Earth
Inventors: Bass; Ronald M., Houston, TX
Assignees: APTI, Inc., Washington, DC
Issued: Mar. 12, 1991
Filed: May. 14, 1987

U.S. Patent 4954709:
High Resolution Directional Gamma Ray Detector
Inventors: Zigler; Arie, Rishon Le Zion, Israel
 Eisen; Yosset, Rishon Le Zion, Israel
Assignees: APTI, Inc., Washington, DC
Issued: Sep. 4, 1990
Filed: Aug. 16, 1989

U.S. Patent 4817495:
Defense System For Discriminating Between Objects In
Space
Inventors: Drobot; Adam T., Annandale, VA
Assignees: APTI, Inc., Los Angeles, CA
Issued: Apr. 4, 1989
Filed: Jul. 7, 1986

U.S. Patent 4873928:
Nuclear-Sized Explosions Without Radiation
Inventors: Lowther; Frank E., Plano, TX
Assignees: APTI, Inc., Los Angeles, CA
Issued: Oct. 17, 1989
Filed: June 15, 1987

APPENDIX C
WEATHER
MODIFICATION PATENTS

Weather control patents go back to the 1890s. Here is a list of some, starting in the 1920s, compiled by Lorie Kramer in 2003.

[Format: patent Number – Date Granted – Title:]

1338343 - April 27, 1920 - Process And Apparatus For The Production Of Intense Artificial Clouds, Fogs, Or Mists
1619183 - March 1, 1927 - Process Of Producing Smoke Clouds From Moving Aircraft
1631753 - June 7, 1927 - Electric Heater - Referenced In 3990987
1665267 - April 10, 1928 - Process Of Producing Artificial Fogs
1892132 - December 27, 1932 - Atomizing Attachment For Airplane Engine Exhausts
1928963 - October 3, 1933 - Electrical System And Method
1957075 - May 1, 1934 - Airplane Spray Equipment
2097581 - November 2, 1937 - Electric Stream Generator - Referenced In 3990987
2409201 - October 15, 1946 - Smoke Producing Mixture
2476171 - July 18, 1945 - Smoke Screen Generator
2480967 - September 6, 1949 - Aerial Discharge Device
2550324 - April 24, 1951 - Process For Controlling Weather
2510867 - October 9, 1951 - Method Of Crystal Formation And Precipitation
2582678 - June 15, 1952 - Material Disseminating Apparatus For Airplanes

2591988 - April 8, 1952 - Production Of Tio2 Pigments - Referenced In 3899144

2614083 - October 14, 1952 - Metal Chloride Screening Smoke Mixture

2633455 - March 31, 1953 - Smoke Generator

2688069 - August 31, 1954 - Steam Generator - Referenced In 3990987

2721495 - October 25, 1955 - Method And Apparatus For Detecting Minute Crystal Forming Particles Suspended In A Gaseous Atmosphere

2730402 - January 10, 1956 - Controllable Dispersal Device

2801322 - July 30, 1957 - Decomposition Chamber For Monopropellant Fuel - Referenced In 3990987

2881335 - April 7, 1959 - Generation Of Electrical Fields

2908442 - October 13, 1959 - Method For Dispersing Natural Atmospheric Fogs And Clouds

2986360 - May 30, 1962 - Aerial Insecticide Dusting Device

2963975 - December 13, 1960 - Cloud Seeding Carbon Dioxide Bullet

3126155 - March 24, 1964 - Silver Iodide Cloud Seeding Generator - Referenced In 3990987

3127107 - March 31, 1964 - Generation Of Ice-Nucleating Crystals

3131131 - April 28, 1964 - Electrostatic Mixing In Microbial Conversions

3174150 - March 16, 1965 - Self-Focusing Antenna System

3234357 - February 8, 1966 - Electrically Heated Smoke Producing Device

3274035 - September 20, 1966 - Metallic Composition For Production Of Hydroscopic Smoke

3300721 - January 24, 1967 - Means For Communication Through A Layer Of Ionized Gases

3313487 - April 11, 1967 - Cloud Seeding Apparatus

3338476 - August 29, 1967 - Heating Device For Use With Aerosol Containers - Referenced In 3990987

3410489 - November 12, 1968 - Automatically Adjustable Airfoil Spray System With Pump

3429507 - February 25, 1969 - Rainmaker
3432208 - November 7, 1967 - Fluidized Particle Dispenser
3441214 - April 29, 1969 - Method And Apparatus For
Seeding Clouds
3445844 - May 20, 1969 - Trapped Electromagnetic
Radiation Communications System
3456880 - July 22, 1969 - Method Of Producing
Precipitation From The Atmosphere
3518670 June 30, 1970 - Artificial Ion Cloud
3534906 - October 20, 1970 - Control Of Atmospheric
Particles
3545677 - December 8, 1970 - Method Of Cloud Seeding
3564253 - February 16, 1971 - System And Method For
Irradiation Of Planet Surface Areas
3587966 - June 28, 1971 - Freezing Nucleation
3601312 - August 24, 1971 - Methods Of Increasing The
Likelihood Of Precipitation By The Artificial Introduction
Of Sea Water Vapor Into The Atmosphere Windward Of An
Air Lift Region
3608810 - September 28, 1971 - Methods Of Treating
Atmospheric Conditions
3608820 - September 20, 1971 - Treatment Of Atmospheric
Conditions By Intermittent Dispensing Of Materials
Therein
3613992 - October 19, 1971 - Weather Modification Method
3630950 - December 28, 1971 - Combustible Compositions
For Generating Aerosols, Particularly Suitable For Cloud
Modification And Weather Control And Aerosolization
Process
USRE29142 - This Patent Is A Reissue Of Patent US3630950
- Combustible Compositions For Generating Aerosols,
Particularly Suitable For Cloud Modification And Weather
Control And Aerosolization Process
3659785 - December 8, 1971 - Weather Modification
Utilizing Microencapsulated Material
3666176 - March 3, 1972 - Solar Temperature Inversion
Device

3677840 - July 18, 1972 - Pyrotechnics Comprising Oxide Of Silver For Weather Modification Use

3722183 - March 27, 1973 - Device For Clearing Impurities From The Atmosphere

3769107 - October 30, 1973 - Pyrotechnic Composition For Generating Lead Based Smoke

3784099 - January 8, 1974 - Air Pollution Control Method

3785557 - January 15, 1974 - Cloud Seeding System

3795626 - March 5, 1974 - Weather Modification Process

3808595 - April 30, 1974 - Chaff Dispensing System

3813875 - June 4, 1974 - Rocket Having Barium Release System To Create Ion Clouds In The Upper Atmosphere

3835059 - September 10, 1974 - Methods Of Generating Ice Nuclei Smoke Particles For Weather Modification And Apparatus Therefore

3835293 - September 10, 1974 - Electrical Heating Apparatus For Generating Super Heated Vapors - Referenced In 3990987

3877642 - April 15, 1975 - Freezing Nucleant

3882393 - May 6, 1975 - Communications System Utilizing Modulation Of The Characteristic Polarization Of The Ionosphere

3896993 - July 29, 1975 - Process For Local Modification Of Fog And Clouds For Triggering Their Precipitation And For Hindering The Development Of Hail Producing Clouds

3899129 - August 12, 1975 - Apparatus For Generating Ice Nuclei Smoke Particles For Weather Modification

3899144 - August 12, 1975 - Powder Contrail Generation

3940059 - February 24, 1976 - Method For Fog Dispersion

3940060 - February 24, 1976 - Vortex Ring Generator

3990987 - November 9, 1976 - Smoke Generator

3992628 - November 16, 1976 - Countermeasure System For Laser Radiation

3994437 - November 30, 1976 - Broadcast Dissemination Of Trace Quantities Of Biologically Active Chemicals

4042196 - August 16, 1977 - Method And Apparatus For

Triggering A Substantial Change In Earth Characteristics And Measuring Earth Changes

RE29,142 - February 22, 1977 - Reissue Of: 03630950 - Combustible Compositions For Generating Aerosols, Particularly Suitable For Cloud Modification And Weather Control And Aerosolization Process

4035726 - July 12, 1977 - Method Of Controlling And/Or Improving High-Latitude And Other Communications Or Radio Wave Surveillance Systems By Partial Control Of Radio Wave Et Al

4096005 - June 20, 1978 - Pyrotechnic Cloud Seeding Composition

4129252 - December 12, 1978 - Method And Apparatus For Production Of Seeding Materials

4141274 - February 27, 1979 - Weather Modification Automatic Cartridge Dispenser

4167008 - September 4, 1979 - Fluid Bed Chaff Dispenser

4347284 - August 31, 1982 - White Cover Sheet Material Capable Of Reflecting Ultraviolet Rays

4362271 - December 7, 1982 - Procedure For The Artificial Modification Of Atmospheric Precipitation As Well As Compounds With A Dimethyl Sulfoxide Base For Use In Carrying Out Said Procedure

4402480 - September 6, 1983 - Atmosphere Modification Satellite

4412654 - November 1, 1983 - Laminar Microjet Atomizer And Method Of Aerial Spraying Of Liquids

4415265 - November 15, 1983 - Method And Apparatus For Aerosol Particle Absorption Spectroscopy

4470544 - September 11, 1984 - Method Of And Means For Weather Modification

4475927 - October 9, 1984 - Bipolar Fog Abatement System

4600147 - July 15, 1986 - Liquid Propane Generator For Cloud Seeding Apparatus

4633714 - January 6, 1987 - Aerosol Particle Charge And Size Analyzer

4643355 - February 17, 1987 - Method And Apparatus For

Modification Of Climatic Conditions
4653690 - March 31, 1987 - Method Of Producing Cumulus Clouds
4684063 - August 4, 1987 - Particulates Generation And Removal
4686605 - August 11, 1987 - Method And Apparatus For Altering A Region In The Earth's Atmosphere, Ionosphere, And/Or Magnetosphere
4704942 - November 10, 1987 - Charged Aerosol
4712155 - December 8, 1987 - Method And Apparatus For Creating An Artificial Electron Cyclotron Heating Region Of Plasma
4744919 - May 17, 1988 - Method Of Dispersing Particulate Aerosol Tracer
4766725 - August 30, 1988 - Method Of Suppressing Formation Of Contrails And Solution Therefore
4829838 - May 16, 1989 - Method And Apparatus For The Measurement Of The Size Of Particles Entrained In A Gas
4836086 - June 6, 1989 - Apparatus And Method For The Mixing And Diffusion Of Warm And Cold Air For Dissolving Fog
4873928 - October 17, 1989 - Nuclear-Sized Explosions Without Radiation
4948257 - August 14, 1990 - Laser Optical Measuring Device And Method For Stabilizing Fringe Pattern Spacing
4948050 - August 14, 1990 - Liquid Atomizing Apparatus For Aerial Spraying
4999637 - March 12, 1991 - Creation Of Artificial Ionization Clouds Above The Earth
5003186 - March 26, 1991 - Stratospheric Welsbach Seeding For Reduction Of Global Warming
5005355 - April 9, 1991 - Method Of Suppressing Formation Of Contrails And Solution Therefore
5038664 - August 13, 1991 - Method For Producing A Shell Of Relativistic Particles At An Altitude Above The Earths Surface
5041760 - August 20, 1991 - Method And Apparatus

For Generating And Utilizing A Compound Plasma Configuration

5041834 - August 20, 1991 - Artificial Ionospheric Mirror Composed Of A Plasma Layer Which Can Be Tilted

5056357 - October 15, 1991- Acoustic Method For Measuring Properties Of A Mobile Medium

5059909 - October 22, 1991 - Determination Of Particle Size And Electrical Charge

5104069 - April 14, 1992 - Apparatus And Method For Ejecting Matter From An Aircraft

5110502 - May 5, 1992 - Method Of Suppressing Formation Of Contrails And Solution Therefore

5156802 - October 20, 1992 - Inspection Of Fuel Particles With Acoustics

5174498 - December 29, 1992 - Cloud Seeding

5148173 - September 15, 1992 - Millimeter Wave Screening Cloud And Method

5245290 - September 14, 1993 - Device For Determining The Size And Charge Of Colloidal Particles By Measuring Electroacoustic Effect

5286979 - February 15, 1994 - Process For Absorbing Ultraviolet Radiation Using Dispersed Melanin

5296910 - March 22, 1994 - Method And Apparatus For Particle Analysis

5327222 - July 5, 1994 - Displacement Information Detecting Apparatus

5357865 - October 25, 1994 - Method Of Cloud Seeding

5360162 - November 1, 1994 - Method And Composition For Precipitation Of Atmospheric Water

5383024 - January 17, 1995 - Optical Wet Steam Monitor

5425413 - June 20, 1995 - Method To Hinder The Formation And To Break-Up Overhead Atmospheric Inversions, Enhance Ground Level Air Circulation And Improve Urban Air Quality

5434667 - July 18, 1995 - Characterization Of Particles By Modulated Dynamic Light Scattering

5441200 - August 15, 1995 - Tropical Cyclone Disruption

5486900 - January 23, 1996 - Measuring Device For Amount Of Charge Of Toner And Image Forming Apparatus Having The Measuring Device

5556029 - September 17, 1996 - Method Of Hydrometeor Dissipation (Clouds)

5628455 - May 13, 1997 - Method And Apparatus For Modification Of Supercooled Fog

5631414 - May 20, 1997 - Method And Device For Remote Diagnostics Of Ocean-Atmosphere System State

5639441 - June 17, 1997 - Methods For Fine Particle Formation

5762298 - June 9, 1998 - Use Of Artificial Satellites In Earth Orbits Adaptively To Modify The Effect That Solar Radiation Would Otherwise Have On Earth's Weather

5912396 - June 15, 1999 - System And Method For Remediation Of Selected Atmospheric Conditions

5922976 - July 13, 1999 - Method Of Measuring Aerosol Particles Using Automated Mobility-Classified Aerosol Detector

5949001 - September 7, 1999 - Method For Aerodynamic Particle Size Analysis

5984239 - November 16, 1999 - Weather Modification By Artificial Satellite

6025402 - February 15, 2000 - Chemical Composition For Effectuating A Reduction Of Visibility Obscuration, And A Detoxification Of Fumes And Chemical Fogs In Spaces Of Fire Origin

6030506 - February 29, 2000 - Preparation Of Independently Generated Highly Reactive Chemical Species

6034073 - March 7, 2000 - Solvent Detergent Emulsions Having Antiviral Activity

6045089 - April 4, 2000 - Solar-Powered Airplane

6056203 - May 2, 2000 - Method And Apparatus For Modifying Supercooled Clouds

6110590 - August 29, 2000 - Synthetically Spun Silk Nanofibers And A Process For Making The Same

6263744 - July 24, 2001 - Automated Mobility-Classified-

Aerosol Detector

6281972 - August 28, 2001 - Method And Apparatus For Measuring Particle-Size Distribution

6315213 - November 13, 2001 - Method Of Modifying Weather

6382526 - May 7, 2002 - Process And Apparatus For The Production Of Nanofibers

6408704 - June 25, 2002 - Aerodynamic Particle Size Analysis Method And Apparatus

6412416 - July 2, 2002 - Propellant-Based Aerosol Generation Devices And Method

6520425 - February 18, 2003 - Process And Apparatus For The Production Of Nanofibers

6539812 - April 1, 2003 - System For Measuring The Flow-Rate Of A Gas By Means Of Ultrasound

6553849 - April 29, 2003 - Electrodynamic Particle Size Analyzer

6569393 - May 27, 2003 - Method And Device For Cleaning The Atmosphere

BIBLIOGRAPHY

BOOKS

Alexander, Colonel John B., US Army (ret). *Future War: Non-Lethal Weapons in Twenty-First Century Warfare.* St. Martin's Press, 1999.

Allen, Gary. *None Dare Call It Conspiracy.* Concord Press, 1972.

Athanasiou, Tom and Paul Baer. *Dead Heat: Global Justice and Global Warming.* Open Media, 2002.

Barnes-Svarney, Patricia and Thomas E. Svarney. *Skies Of Fury: Weather Weirdness Around the World.* A Touchstone Book, Published by Simon & Schuster, 1999.

Battan, Louis J. *Harvesting the Clouds: Advances in Weather Modification.* Doubleday, 1969.

Bearden, Lt. Col. Thomas E. (ret). *Fer de Lance: A Briefing On Soviet Scalar Electromagnetic Weapons.* Tesla Book Company, 1986.

Begich, Nick, and Jeane Manning. *Angels Don't Play This HAARP: Advances In Tesla Technology.* Earthpulse Press, 1995.

Begich, Nick and James Roderick. *Earth Rising: The Revolution – Toward a Thousand Years of Peace.* Earthpulse Press, 2000.

Begich, Nick and James Roderick. *Earth Rising II: The Betrayal of Science, Society and the Soul.* Earthpulse Press, 2003.

Bell, Art and Whitley Strieber. *The Coming Global Superstorm.* Pocket Books, 2000.

Bertell, Rosalie. *No Immediate Danger: Prognosis for a Radioactive Earth.* Book Publishing Company, 2000.

Bertell, Rosalie. *Planet Earth: The Latest Weapon of War.* The Women's Press Ltd., 2000.

351

Weather Warfare

Bowart, Walter. *Operation: Mind Control*. Dell Publishing, 1978.

Boyle, Francis A. *The Criminality of Nuclear Deterrence: Could The U.S. War On Terrorism Go Nuclear?* Clarity Press, 2002.

Breuer, Georg, Hans Morth (Translator). *Weather Modification: Prospects and Problems*. Cambridge University Press, 1980.

Brezinski, Zbigniew. *Between Two Ages: America's Role in the Technetronic Era. Penguin Books, 1976.*

Brown, Thomas J. *Loom of the Future: The Weather Engineering Work of Trevor James Constable*. Borderland Sciences, 1994.

Carson, Rachel. *Silent Spring*. Houghton Mifflin Co., 1962.

Clarke, Lee. *Worst Cases: Terror and Catastrophe in the Popular Imagination*. University of Chicago Press, 2005.

Coble, Barry B. *Benign Weather Modification*, School of Advanced Airpower Studies, March, 1997.

Cole, Leonard A. *Clouds of Secrecy, The Army's Germ Warfare Tests Over Populated Areas*. Rowman & Littlefield, 1988.

Congressional Research Service. *Weather Modification*. University Press of the Pacific, 2004.

Console, Rodolfo and Alexei Nikolaev (Editors). *Earthquakes Induced by Nuclear Explosions: Environmental and Ecological Problems* (NATO ASI Series. Series 2, Environment, Vol 4). Springer-Verlag Telos, 1995.

Constable, James Trevor. *Cosmic Pulse of Life*. Ariel Press, 1977.

Constantine, Alex. *Virtual Government*. Feral House, 1997.

Corrington, Robert S. *Wilhelm Reich: Psychoanalyst and Radical Naturalist*. Farrar, Straus and Giroux, 2003.

Cotton, William R. and Roger A. Pielke. *Human Impacts on Weather and Climate*. Cambridge University Press, 1995.

Cox, John D. *Weather for Dummies*. For Dummies, 2000.

Cox, John D. *Storm Watchers: The Turbulent History of Weather Prediction from Franklin's Kite to El Nino*. Wiley, 2002.

DeMeo, James, foreword by Eva Reich. *The Orgone Accumulator Handbook: Construction Plans Experimental Use and Protection Against Toxic Energy*. Natural Energy Works, 1989.

Dotto, Lydia. *The Ozone War*. Doubleday, 1978.

Dunn, Christopher. *The Giza Power Plant: Technologies of Ancient Egypt*. Bear & Co., 1998.

Erikson, Kai. *A New Species of Trouble*. W.W. Norton & Co., 1994.

Farrell, Joseph P. *The Giza Death Star*. Adventures Unlimited Press, 2001.

Farrell, Joseph P. *The Giza Death Star Deployed: The Physics and Engineering of the Great Pyramid*. Adventures Unlimited Press, 2003.

Farrell, Joseph P. *The Giza Death Star Destroyed: The Ancient War For Future Science*. Adventures Unlimited Press, 2001.

Felix, Robert W. *Not by Fire but by Ice: Discover What Killed the Dinosaurs... and Why It Could Soon Kill Us*. Sugarhouse Publishing, 1999.

Flannery, Tim. *The Weather Makers: How Man Is Changing the Climate and What It Means for Life on Earth*. Atlantic Monthly Press, 2006.

Fleagle, Robert G. (Ed). *Weather Modification in Public Interest*. University of Washington Press, 1974.

Fleagle, Robert G. *Weather Modification: Science and Public Policy*. University of Wisconsin Press, 1969.

Foreign Policy Association. *Toward the Year 2018*. Cowles Education Corp., 1968.

Gelbspan, Ross. *The Heat Is on: The Climate Crisis, the Cover-Up, the Prescription.* Perseus Books Group, 1998.

Gribbin, John R. and Stephen H. Plagemann. *The Jupiter Effect: The Planets As Triggers Of Devastating Earthquakes.* Vintage Books, 1976.

Halacy, Jr., Daniel S. *The Weather Changers.* Harper & Row, 1968.

Havens, Barrington S., James E. Jiusto, Bernard Vonnegut. *Early History of Cloud Seeding.* 75 pp., booklet jointly published by Langmuir Laboratory at the New Mexico Institute of Mining and Technology, Atmospheric Sciences Research Center at the State University of New York at Albany, and Research and Development Center of General Electric Company, 1978. This booklet includes annotations by Profs. Vonnegut and Jiusto, together with a reprint of: Barrington S. Havens, "History of Project Cirrus," *General Electric Research Laboratory Report,* RL-756, 1952. A condensed version of Havens' report was published in *General Electric Review,* Vol. 55, pp. 8-26, November 1952.

Hess, W. N. (Ed). *Weather and Climate Modification,* Wiley-Interscience, 1974.

Horowitz, Len, Dr., *Death in the Air: Globalism, Terrorism & Toxic Warfare.* Tetrahedron, 2001.

Houghton, John T., L. G. Meiro Filho, B. A. Callander, N. Harris, A. Kattenburg, K. Maskell (Editors). *Climate Change 1995: The Science of Climate Change: Contribution of Working Group I to the Second Assessment Report of the Intergovernmental Panel on Climate Change.* Cambridge University Press, 1996.

Houghton, John. *Global Warming: The Complete Briefing.* Cambridge University Press, 2004.

James, G. E. *Chaos Theory: The Essentials for Military Applications,* ACSC Theater Air Campaign Studies Course Book, AY96, Vol. 8. Maxwell AFB, Ala.: Air University Press, 1995.

Jewell, Roger L. *Riding The Wild Orb: Long-Term Weather Extremes On*

The Planet. Jewell Histories, 2001.

Johnstone, A.K., Ph.D. *UFO Defense Tactics: Weather Shield to Chemtrails.* Hancock House, 2002.

Keith, Jim. *Mind Control, World Control: The Encyclopedia of Mind Control.* Adventures Unlimited Press, 1997.

Kelley, Charles R. *LIFE FORCE: The Creative Process in Man and in Nature.* Trafford Publishers, 2004.

Kolbert, Elizabeth. *Field Notes from a Catastrophe.* Bloomsbury USA, 2006.

Kondratyev, K.Ya. *Climatic Effects of Aerosols and Clouds.* Springer Praxis Books, 1999.

Larranaga, Jim. *The Dead Farmer's Almanac: Who Really Controls the Weather?* Xlibris Corporation, 2000.

Leggett, Jeremy K. *The Carbon War: Global Warming and the End of the Oil Era.* Routledge, 2001.

Linden, Eugene. *The Winds of Change: Climate, Weather, and the Destruction of Civilizations.* Simon & Schuster, 2006.

Lockhart, Gary. *The Weather Companion: An Album of Meteorological History, Science, and Folklore.* Wiley, 1988.

Lovelock, J. E. *Gaia: A New Look at Life on Earth,* Oxford University Press, 1979.

Lovelock, James. *Healing Gaia: Practical Medicine for the Planet,* Harmony Books, 1991.

Ludlum, David. *National Audubon Society Field Guide to North American Weather.* Knopf, 1991.

Malone, Thomas F. *Weather and Climate Modification: Problems and Progress.* Gale Group, 1980.

Weather Warfare

Miller, Clark, Paul N. Edwards (Editors). *Changing the Atmosphere: Expert Knowledge and Environmental Governance*. MIT Press, 2001.

Miller, Judith, William Broad, Stephen Engelberg. *Germs: Biological Weapons and America's Secret War*. Simon & Schuster, 2002.

Moreno, Jonathan D. *Undue Risk, Secret State Experiments on Humans*. Freeman & Co., 1999.

National Research Council of the National Academies of Sciences. *Critical Issues in Weather Modification Research*. National Academies Press, 2003.

National Research Council of the National Academies of Sciences. *Weather & Climate Modification: Problems And Progress*. National Academies Press, 1973.

Pellegrino, Charles. *Unearthing Atlantis: An Archaeological Odyssey to the Fabled Lost Civilization*. Avon, 2001.

Peters, R.L. and T.E. Lovejoy (Eds), *Global Warming and Biological Diversity*. Yale University Press, 1994.

Philander, S. George. *Is the Temperature Rising? The Uncertain Science of Global Warming*. Princeton University Press, 2000.

Pond, Dale, Walter Baumgartner. *Nikola Tesla's Earthquake Machine: With Tesla's Original Patents Plus New Blueprints to Build Your Own Working Model*. Message Company, 1995.

Ponte, Lowell. *The Cooling*. Prentice-Hall, Inc., 1976.

Posner, Richard. *Catastrophe: Risk and Response*. Oxford University Press, 2004.

Regis, Ed. *The Biology of Doom: America's Secret Germ Warfare Project*. Owl Books, 2000.

Schneider, Stephen H. *Climate Change Policy: A Survey*. Island Press, 2002.

Semler, Eric, James Benjamin and Adam Gross; Sarah Rosenfield, Senior Researcher. *The Language of Nuclear War: An Intelligent Citizen's Dictionary*. Harper & Row, 1987.

Sewell, Derrick W. R. *Human Dimensions of Weather Modification* (Research Papers Series No. 105). University of Chicago, 1966.

Shukman, David. *Tomorrow's War: The Threat of High-technology Weapons*. Harcourt Brace & Company, 1996.

Siegel, Lenny. *Operation Ozone Shield: The Pentagon's War on the Stratosphere*. The Fund, 1992.

Stucki, Margaret E. *Croak! USSR-USA Top Secret Weather Control Tesla-Tech (2nd Edition of Elegies in Ecology or Eco-Elegia)*. Duverus Publishing Corp., 2005.

Swartz, Tim and Timothy Beckley. *The Lost Journals of Nikola Tesla: HAARP - Chemtrails and Secret of Alternative 4*. Inner Light - Global Communications, 2000.

Taubenfeld, Howard J. (Ed). *Controlling the Weather: A Study of Law and Regulatory Procedures*. Dunellen Co., 1970.

Tesla, Nikola and David Hatcher Childress. *The Fantastic Inventions of Nikola Tesla*. Adventures Unlimited Press, 1993.

Tesla, Nikola and David Hatcher Childress (Ed). *The Tesla Papers: Nikola Tesla on Free Energy & Wireless Transmission of Power*. Adventures Unlimited Press, 2000.

Thomas, William. *Chemtrails Confirmed*. Bridger House Publishers, 2004.

Thomas, William A. (Ed). *Legal and Scientific Uncertainties of Weather Modification*. Duke University Press, 1977.

Trinkaus, George. *Tesla: The Lost Inventions*. High Voltage Press, 1988.

Tsarion, Michael. *Atlantis, Alien Visitation & Genetic Manipulation*. Taroscopes, 2002.

Turney, Jon. *Lovelock & Gaia: Signs of Life*. Columbia University Press. 2003.

Upgren, Arthur and Jurgen Stock. *Weather: How It Works and Why It Matters*. Perseus Publishing, 2000.

Valone, Thomas, Ph.D., PE. *Harnessing The Wheelwork Of Nature: Tesla's Science Of Energy*. Adventures Unlimited Press, 2002.

Vassilatos, Gerry. *Secrets of Cold War Technology: Project HAARP and Beyond*. Adventures Unlimited Press, 2000.

Watts, Robert G. *Response to Global Climate Change: Planning a Research and Development Agenda*. CRC, 1997.

Weart, Spencer R. *The Discovery of Global Warming*. Harvard University Press, 2004.

Westing, Arthur H. (Ed). *Environmental Warfare: A Technical, Legal and Policy Appraisal*. Taylor & Francis, 1984.

Williams, Jack. *The USA Today Weather Book: An Easy-To-Understand Guide to the USA's Weather*. Vintage, 1997.

Wilson, R.; M. Zinyowera; R. Moss (Eds). *Climate Change 1995: Impacts, Adaptations and Mitigation of Climate Change: Scientific-Technical Analyses*. Cambridge University Press 1996.

Winters, Harold A., Gerald A. Galloway, William J. Reynolds, David W. Rhyne. *Battling the Elements: Weather and Terrain in the Conduct of War*. The Johns Hopkins University Press, 2001.

ARTICLES, NEWS STORIES & SCIENTIFIC REPORTS

Aber, J.D., K.J. Nadelhoffer, P. Steudler, and J. Mellilo, "Nitrogen Saturation In Northern Forest Ecosystems." *Bioscience*, 39 (6), 378-386. 1989.

Aditjondro, G. "The Driving Force of Indonesia's Catastrophic Forest Fires." *Eco-Politics Journal*, October 2000.

Air University of the US Air Force. "AF 2025 Final Report." http://csat.au.af.mil/2025/index.htm

Alaska North Slope Electric Missile Shield, Defense Advanced Research Projects Agency (DARPA), Contract No. DAAHD1-86-C-0420, February 1986, Arlington, VA.

Alexander, Verne. "The Greatest Flood of History," *Weatherwise*, Vol. 4, pp. 110-111, October ,1951.

American Geophysical Union, "Jet contrails to be significant climate factor by 2050," *AGU PRESS RELEASE* NO. 99-19, June 21, 1999.

American Meteorological Society, "Scientific Background for the AMS Policy Statement on Planned and Inadvertent Weather Modification," *Bulletin of the American Meteorological Society*, Vol. 79, pp. 2773-78, December, 1998.

Ames, R.B., and W.C. Malm, "Estimating the Contribution of the Mohave Coal-fired Power Plant Emissions to Atmospheric Extinction at Grand Canyon National Park." In *Visual Air Quality: Aerosols and Global Radiation Balance*, Air & Waste Management Association, Pittsburgh, PA., 683-697, 1997.

Appleman, H. "An Introduction to Weather-modification." Scott AFB, Ill.: *Air Weather Service*.

Appleman, H. "The formation of exhaust condensation trails by jet aircraft. "*Bull. Amer. Meteor. Soc.*, 1953.

Arnott, W. P., Y.Y. Dong, J. Hallett and M. R. Poellot, "Observations and importance of small ice crystals in a cirrus cloud from FIRE II data." *J. Geophys*. Res., 99, 1371-1381, 1994.

Arnott, W.P., Y. Dong, R. Purcell and J. Hallett, "Direct airborne sampling of small ice crystals and the concentration and phase of haze particles." 9th Sym. on Meteorological Observations and Instrumentation. Charlotte, North Carolina, 415-420. 1995.

Bailey, Patrick, Nancy Worthington. "HISTORY AND APPLICATIONS OF HAARP TECHNOLOGIES: THE HIGH FREQUENCY ACTIVE AURORAL RESEARCH PROGRAM." An updated and revised version of a paper originally submitted in June 1997 to the 1997 Intersociety Energy Conversion Engineering Conference, the 32nd IECEC, held July 27 - August 1, 1997, in Honolulu, Hawaii. http://www.padrak.com/ine/HAARP97.html

Ball, Vaughn C. "Shaping the Law of Weather Control," *The Yale Law Journal,* Vol. 58, pp. 213-244, January 1949.

Banks, Peter M. "Overview of Ionospheric Modification from Space Platforms, in Ionospheric Modification and Its Potential to Enhance or Degrade the Performance of Military Systems," *AGARD Conference Proceedings 485*, October 1990.

A. Barnes, "Weather Modification," Proceedings of TECOM Test Technology Symposium, 97, U.S. Army Test and Evaluation Command, March 18, 1997.

Baughcum, S. L. "Subsonic aircraft emission inventories, In: *Atmospheric Effects of Aviation: First Report of the Subsonic Assessment Project." NASA RP-1385*, pp. 15-29, 1996.

Begich, Nick. "Biohazards of Extremely Low Frequencies (ELF)Biohazards of Extremely Low Frequencies (ELF)." *Earthpulse Flashpoints* Series 1 Volume 1.

Begich, Nick, Jeane Manning. "Ground-Based 'Star Wars' – Disaster Or 'Pure' Research?" *Earthpulse Flashpoints*, Newtext Number Three.

Begich, Nick, Jeane Manning. "Vandalism In The Sky?" *Nexus Magazine*, Volume 3, Number 1 (December '95-January '96).

Benford, Gregory. "Climate Controls." *Reason*, November 1997.

Bertell, Rosalie. "Background of the HAARP Program." *Earthpulse Press* November 5, 1996.

Bingham, Eugene. "Devastating tsunami bomb viable, say experts."

The New Zealand Herald, September 28, 1999.

Bingham, Eugene. "Tsunami bomb NZ's devastating war secret." *The New Zealand Herald*, September 25, 1999.

Bodily, Samuel E., Robert F. Bruner, and Mari Capestany. "Enron Corporation's Weather Derivatives (A) & (B)" (2000). Darden Case Nos.: UVA-F-1299-M-SSRN and UVA-F-1300-M-SSRN. Available at SSRN: http://ssrn.com/abstract=274195.

Black, R. A. and J. Hallett, "The Mystery of Cloud Electrification." *American Scientist*, 86, 526-534, 1998.

Black, R. A. and J. Hallett, "Electrification of the Hurricane." *J. Atmos. Sci.*, 56, 2004-2028, 1999.

Brackman, S. "Bomb Tests and Earthquakes" *War and Peace Digest*, Vol 2, No. 3 Aug. 1992.

Brown, Thomas. "Radionics: At the Crossroads of Science & Magic." http://www.altered-states.co.nz/index2.htm?/radionic/radio.htm

Caldwell, Martha B. "Some Kansas Rain Makers." *Kansas Historical Quarterly,* August, 1938 (Vol. 7, No. 3), pages 306 to 324.

Carleton, Andrew M. "Climatology: Contrails Reduce Daily Temperature Range." *Nature.* August 8, 2002.

Centner, Christopher et al. "Environmental Warfare: Implications for Policymakers and War Planners." Maxwell AFB, Ala.: Air Command and Staff College, May 1995.

Pimiento Chamorro, Susana, Edward Hammond. "Addressing Environmental Modification in Post-Cold War Conflict -- The Convention on the Prohibition of Military or Any other Hostile Use of Environmental Modification Techniques (ENMOD) and Related Agreements. An occasional paper of *The Edmonds Institute*, 2001.

Changnon, S.A. "Midwestern Cloud, Sunshine And Temperature Records Since 1901: Possible Evidence Of Jet Contrail Effects." *Journal of Applied Meteorology*, 1981.

Chossudovsky, Michael. "Washington's New World Order Weapons Have the Ability to Trigger Climate Change." *Center for Research on Globalization*, 2002. http://www.globalresearch.ca/articles/CHO201A.html.

Coble, Barry B. "Benign Weather Modification." *School of Advanced Airpower Studies*, March, 1997.

"Commerce Scientists Heat the Ionosphere." *United States Department of Commerce News* 7 Dec. 1970.

Cook, III. Maj. Joseph W. et al. Nonlethal Weapons: Technologies, Legalities, and Potential Policies. *Airpower Journal*, 9, (SE) 77-91. [177] 1994.

Cook-Anderson, Gretchen. "Clouds Caused By Aircraft Exhaust May Warm The U.S. Climate," *NASA PRESS RELEASE* 04-140, Langley Research Center, Hampton, Va. April 27, 2004.

Corrales, Scott. "Sky Pilots of the Apocalypse: Weather As a Weapon." *Fate*, Vol. 54, No. 7, July 2001.

Cotton, William R. "Testing, Implementation, and Evolution of Seeding Concepts — A Review," pp. 139-149, in Roscoe R. Braham, Jr., editor, *Precipitation Enhancement — A Scientific Challenge*, Meteorological Monographs, Vol. 21, Nr. 43, December 1986.

Cotton, William R. "Weather Modification by Cloud Seeding — A Status Report 1989-1997." Colorado State University, Department of Atmospheric Science. http://rams.atmos.colostate.edu/gkss.html

CNN.com "Mongolian storm pollutes North America." April 17, 2001.

Dalton, Marcus K. "Chemtrails Are Over Las Vegas." *Las Vegas Tribune*, Volume 7 Issue 3, August 19, 2005.

DeGrand, J.Q., A.M. Carleton, D.J., Travis, and P. Lamb. "A Satellite Based Climatic Description Of Jet Aircraft Contrails And Associations With Atmospheric Conditions, 1977-79." *Journal of*

Applied Meteorology, Vol. 39, 2000

DeMaria, M. and J. Kaplan "An Updated Statistical Hurricane Intensity Prediction Scheme (SHIPS) for the Atlantic and Eastern North Pacific Basins" *Weather and Forecasting* vol. 14 no.3 pp. 326-337, 1999.

DeMeo, James. "OROP ARIZONA 1989: A Cloudbusting Experiment To Bring Rains in the Desert Southwest." *Pulse of the Planet #3,* Summer 1991, p.82-92

Drobot, A., et al., Physics Studies in Artificial Ionospheric Mirror (AIM) Related Phenomena, Report GL-TR-90-0038, Geophysics Laboratory, Air Force Systems Command, Hanscom Air Force Base, MA, February 23, 1990.

Drumbl, M. "International Human Rights, International Humanitarian Law, and Environmental Security: Can the International Criminal Court Bridge the Gaps?" *ILSA Journal of International & Comparative Law,* v.6:2, 2000.

Duda, D.P., P. Minnis, and L. Nguyen. "Estimates Of Cloud Radiative Forcing In Contrail Clusters Using GOES Imagery." *Journal of Geophysical Research* 2001.

Duncan, Lewis M. and Robert L. Showen, "Review of Soviet Ionospheric Modification Research," *Ionospheric Modification and Its Potential to Enhance or Degrade the Performance of Military Systems AGARD Conference Proceedings 485,* October 1990.

Eastlund, Bernard. "Systems Considerations of Weather Modifications Experiments Using High Power Electromagnetic Radiation," *Proceedings of Workshop on Space Exploration and Resource Exploitation, EXPLOSPACE,* October 1998.

Eastlund, Bernard. "Mesoscale Diagnostic Requirements for the Thunderstorm Solar Power Satellite Concept," *Proceedings of the 2nd Conference in the Applications of Remote Sensing and GIS for Disaster Management,* January 19-21, 1999, George Washington University, sponsored by NASA and FEMA.

Farmer, Mark. "Mystery in Alaska." *Popular Science* (cover story), September 1995.

Fisher, Drew, Major (Ed). "Presentations from the HAARP Workshop on Ionospheric Heating Diagnostics," 30 April - 2 May 1991, Phillips Laboratory (AFSC), Hanscom AFB, MA 01731-5000, 22 Oct 1991.

Fitrakis, Bob. "When the Army Owns the Weather." *Columbus Alive*, February 5, 2002.

Fraser-Smith, A. C.; C. C. Teague. "Low Frequency Radio Research at Thule, Greenland." STANFORD UNIV CA SPACE TELECOMMUNICATIONS AND RADIOSCIENCE LAB. http://www.stormingmedia.us/19/1990/A199063.html

Friedl, R. R. (ed.). "Atmospheric Effects of Subsonic Aircraft: Interim Assessment Report of the Advanced Subsonic Technology Program." *NASA RP 1400*, 168 pp., 1997.

Frisby, E. M. "Weather-modification in Southeast Asia, 1966-1972." *The Journal of Weather Modification* Vol. 14, No. 1 (April 1982).

Fulghum, David A. "USAF Acknowledges Beam Weapon Readiness." *Aviation Week & Space Technology*, October 4, 2002.

Gallagher, Bill. "AgriTerrorism, Atmospheric Engineering, And The Bush Crime Family." *Los Angeles Independent Media Center*, May 2004.

Gallagher, Bill. "Chemical Aurora Keyhole Surveillance." *Arizona Indymedia*, May 2004.

Garber, Donald P., Minnis, Patrick, and Costulis, Kay P. "A USA Commercial Flight Track Database For Upper Tropospheric Aircraft Emission Studies. European Conference On Aviation, Atmosphere, And Climate" Atmospheric Sciences, NASA Langley Research Center, Hampton, Virginia, USA. http://ntrs.nasa.gov/

Gardner, R.M. (ed.). "ANCAT/EC Aircraft Emissions Inventory for

1991/92 and 2015. Final Report EUR-18179, ANCAT/EC Working Group." Defence Evaluation and Research Agency, Farnborough, United Kingdom, 108 pp., 1998.

Gelt, Joe. "Weather Modification: A Water Resource Strategy to be Researched, Tested Before Tried." *Arroyo*, Spring 1992, Volume 6, No. 1

Gentry, R.C. 1980 "History of Hurricane Research in the U.S. with Special Emphasis on NHRL and Assoc. Groups" *Proceedings of the 13th AMS Tech. Conf. on Hurricanes and Trop. Met.* pp. 6-165.

GlobalSecurity.org. "HAARP Detection and Imaging of Underground Structures Using ELF/VLF Radio Waves." http://www.globalsecurity.org/intell/systems/haarp.htm

Goncalves, Eduardo. "Sick Century." *The Ecologist*, November 22, 2001.

Goncalves, Eduardo. "The Secret Nuclear War." *The Ecologist,* March 22, 2001.

Gothe, M.B. and H. Grassl. "Satellite Remote Sensing Of The Optical Depth And Mean Crystal Size Of Thin Cirrus And Contrails." *Theoretical and Applied Climatology,* 1993.

Graham-Rowe, Duncan. "High Flyers are Scourge of the Skies." *New Scientist.* Vol. 176, Issue 2365, October 19, 2002.

HAARP Research & Applications – Executive Summary, Technical Information Division, Naval Research Laboratory, Washington, DC, June, 1995

Hackworth, Martin. "Does Science Support Idea Of 'Weather Wars?': Former Weatherman's Claim Doesn't Survive Scrutiny." *Idaho State Journal,* May 14, 2006.

Haliburton, Mary-Sue. "Weather Modification a Long-Established, Though Secretive, Reality." *Pure Energy Systems News,* Sept. 6, 2005.

Henderson, Mark. "Scientists Claim Final Proof Of Global

Warming." *The Times*, 6 May 2004

Herskovitz, Don. "Killing Them Softly." *Journal of Electronic Defense*, August 1993.

Hickman, Martin. "Cheap Flights Threaten UK Targets For Carbon Emissions." *The Independent*, January 28, 2006.

Hoag, Philip L. "Weather Modification." *Yellowstone River Publishing*, 1998.

Hoffman, Ross N. "Controlling Hurricanes: Can Hurricanes And Other Severe Tropical Storms Be Moderated Or Deflected?" *Scientific American*, October 2004.

Hoffman, Ross N. "Controlling the Global Weather." *Bulletin of the American Meteorological Society*, Vol. 83, No. 2, pages 241–248; February 2002.

Houghton, Henry G. "Present Position and Future Possibilities of Weather Control," *Bulletin of the American Meteorological Society*, Vol. 38, pp. 567-570, December 1957.

Hume, Capt Edward E. Jr., "Atmospheric and Space Environmental Research Programs in Brazil (U), March 1993." Foreign Aerospace Science and Technology Center, Air Force Intelligence Command, 24 September 1992.

Hunten, D. M. "Residence Times Of Aerosols And Gases In The Stratosphere." *Geophysical Research Letters* 2(1):26ˇ27, 1975.

Huschke, Ralph E. "A Brief History of Weather Modification Since 1946," *Bulletin of the American Meteorological Society*, Vol. 44, pp. 425-429, July 1963.

Interfax News Agency. "U.S. HAARP Weapon Development Concerns Russian Duma." August 10, 2002.

Jackson, Artie. "Statistical Contrail Forecasting." *Journal of Applied Meteorology*, Vol. 40, Issue 2, February 2001.

Jeffrey, Jason. "Earthquakes: Natural or Man-Made?" *New Dawn* No. 89, March-April 2005.

Jin, Ryu. "NK Missile Warhead Found in Alaska." *The Korea Times*, 4 February 2003.

Johnson, Mike, Capt. "Upper Atmospheric Research and Modification-Former Soviet Union (U), supporting document DST-18205-475-92," Foreign Aerospace Science and Technology Center, AF Intelligence Command, 24 September 1992.

Juda, L. "Negotiating A Treaty On Environmental Modification Warfare: The Convention On Environmental Warfare And Its Impact On The Arms Control Negotiations," *International Organization*, v. 32:4, p. 975-991, 1978.

Karl, T.R., P.D. Jones, R.W. Knight, G. Kukla, N. Plummer, V. Razuvayev, K.P. Gallo, J. Lindseay, R. J. Charlson, and T.C. Peterson. "Asymmetric Trends Of Daily Maximum And Minimum Temperature." Bull Amer. Met. Soc., 1993.

Kelly, Laurie A. "Increased Illness Linked To Mystery Powder: More Mountain Patients Reporting Sickness Since Yellow Dust Appeared." *The Alpenhorn News*, February 2006.

Kelly, Laurie A. "Mysterious powder shrouds area." *The Alpenhorn News*, February 2006.

Kossey, Paul A., et al., "Artificial Ionospheric Mirrors (AIM): Concept and Issues," *Ionospheric Modification and its Potential to Enhance or Degrade the Performance of Military Systems, AGARD Conference Proceedings 485*, October 1990.

Langmuir, Irving. "The Growth of Particles in Smokes and Clouds ...," *Proceedings of the American Philosophical Society,* Vol. 92, pp. 167-198, July 1948. Reprinted in *The Collected Works of Irving Langmuir,* Vol. 10, pp. 145-173, 1961.

Langmuir, Irving. "Cloud Seeding by Means of Dry Ice, Silver Iodide, and Sodium Chloride," *Transactions of the New York Academy of Sciences,* Vol. 14, pp. 40-44, November 1951. Reprinted in *The Collected Works of Irving Langmuir,* Vol. 10, pp. 189-196, 1961.

Langmuir, Irving. "Final Report of Project Cirrus, Part II," General Electric Research Laboratory Report, RL-785, May 1953. Reprinted in *The Collected Works of Irving Langmuir,* Vol. 11, 1961.

Lambakis, Steven, James Kiras, Kristin Kolet, With Contributions By: Kathleen Bailey, Colin Gray, Willis Stanley and Robert Turner. "Understanding "Asymmetric" Threats to the United States." *National Institute for Public Policy*, September 2002. www.nipp.org

Leyser, T.B., P.A. Bernhardt, and F.T. Djuth. "Nonlinear Plasma Processes Studies by Electromagnetic HF Pumping of the Ionospheric F Region." *Review of Radio Science 1996-1999*, W. Ross Stone, Ed., Oxford Univ. Press, 1999.

Jeske, H. "Modification Of Tropospheric Propagation Conditions." *Hamburg Univ. Meteorological Inst.* 01 Oct 1990

MacDonald, Gordon J. F. "Geophysical Warfare: How to Wreak the Environment." In: *Unless Peace Comes: A Scientific Forecast of New Weapons*, Calder, Nigel (Ed), Penguin 1968.

Mackedon, Michon. "Project Faultless: Central Nevada's Near Miss as an Atomic Proving Ground." Eureka County Yucca Mountain Information Office. http://www.yuccamountain.org/

Machos, Greg. "Early Attempts To Control Hurricanes." *http://www.hurricaneville.com/project_stormfury.html*

McDonald, James E. "The Physics of Cloud Modification," *Advances in Geophysics*, Vol. 15, pp. 223-303, 1958.

McDonald, James E. "An Historical Note on an Early Cloud-Modification Experiment," *Bulletin of the American Meteorological Society,* Vol. 42, p. 195, March 1961.

Meltz, G. and F.W. Perkins. "Ionospheric Modification Theory: Past, Present And Future." *Radio Science*, 9(11), 1974.

Metwally, M. "Jet Aircraft Engine Emissions Database

368

Development--1992 Military, Charter, and Non-scheduled Traffic."
NASA CR-4684, 1995.

Michaelson, Jay. Geoengineering: "A Climate Change Manhattan
Project." *Stanford Environmental Law Journal*, January, 1998.

Michrowski, Andrew and Peter Webb. "Radionics Case Study: Rain for
Gujarat State, India, 2004." *USPA PROCEEDINGS*, 2005, pg.# 258-263.

Minchin, Rod. "Coral Reefs And Marine Life May Be Wiped Out By
Global Warming." *The Scotsman*, 21 May 2006.

Minnis, P., J.K. Ayers, and S.P. Weaver. "Surface-based Observations
Of Contrail Occurrence Frequency Over The U.S., April 1993 - April
1994." *NASA Ref. Publ. 1404*, May 1997.

Minnis, Patrick. "Contrail Frequency over the United States from
Surface Observations." *Atmospheric Sciences Research*. NASA Langley
Research Center, Hampton, VA. August 12, 2002.

Mintz, John. "Pentagon Fights Secret Scenario Speculation Over
Alaska Antennas." *The Washington Post*, April 17, 1995, pA3.

Mook, Conrad P., Eugene W. Hoover, and Robert A Hoover, "An
Analysis of the Movement of a Hurricane Off the East Coast of the
United States, October 12-14, 1947," *Monthly Weather Review*, Vol. 85,
pp. 243-250, July 1957.

Monastersky, Richard. "Ten Thousand Cloud Makers: Is Airplane
Exhaust Altering Earth's Climate?" *Science News*. Volume 150,
Number 1, July 6, 1996.

Moss, Nan and David Corbin. "Shamanism and the Spirits of
Weather." *Shamanism*, Fall/Winter 1999, Vol. 12, No. 2.

Mueller, A. C., and D. J. Kessler. "The Effects Of Particulates From
Solid Rocket Motors Fired In Space." *Advances in Space Research*
5(2):77ʹ86, 1985.

Murcray, W.B. "On The Possibility Of Weather Modification By
Aircraft Contrails." *Monthly Weather Review*, 1970.

Nakamura, Karen. "Abnormal Phenomena With Nuclear Tests, Earthquakes And The Atmosphere." *The Coastal Post*, August, 1996.

Norris, Michele. "Weather As A Weapon: Manipulating the Weather," *ABCNEWS.com*, February 17, 2002.

Observer, The. "What is the real price of cheap air travel?" January 29, 2006.

Onkow, Nick. "Contrails: What's Left Behind Is Bad News." March 4, 2006.
http://www.airliners.net/articles/read.main?id=85

Orville, H.T., Capt. USN. "Weather Made To Order?" *Collier's*, May 30, 1954.

Page, Douglas. "Climate Change Out of the Blue" *Wired Magazine*, May 10, 2004.

Penson, Stuart. "Europe Emissions Scheme Would Hit Air Fares – BA." *REUTERS*, February 10, 2006.

Perlman, Michael. "Climatic Mayhem and the Hope of the Earth." *American PIE* (Public Information on the Environment), Vol. 3, No. 4, Fall 1997.

Ponte, L. "Pentagon & Kremlin Are Playing With Our Weather And Giving Us Storms and Floods." *Star Magazine*, July, 1982.

Port, Otis. "Rainmaking Has Its True Believers — And Skeptics." *BusinessWeek*, October 24, 2005.

Purrett, Louise A. "Weather Modification As A Future Weapon: The Military May Have Added The Weather To Its Arsenal." *Science News*, April 15, 1972, Vol. 101.

Ramanathan, V. "The Greenhouse Theory Of Climate Change: A Test By An Inadvertent Experiment." *Science*, 1988.

Ramaswamy, V., and J. T. Kiehl. "Sensitivities Of The Radiative

Forcing Due To Large Loadings Of Smoke And Dust Aerosols." *Journal of Geophysical Research*, 90(D3):5597˜5613, 1985.

Rembert, Tracey C. "Discordant HAARP." *E, The Environmental Magazine*, January 1997.

Rincon, Paul. "Telescopes 'Worthless' By 2050." *BBC News*, 2 March 2006.

Rosenberg, Eric. "New Face of Terrorism: Radio-Frequency Weapons." *Hearst Newspapers*, 1997.

Rosenfeld, Albert. "The Quintessence of Irving Langmuir: A Biography," published in *The Collected Works of Irving Langmuir*, Vol. 12, 1961.

Sand, P. "International Environmental Law after Rio," *European Journal of International Law*, v. 4:3, 1993.

Sassen, K. "Contrail Cirrus And Their Potential For Regional Climate Change." *Bull. Amer. Meteor. Soc.*, Vol. 78, No. 9, 1997.

Seagraves, Mary Ann, Richard Szymber. "Weather a Force Multiplier" *Military Review*, November/December 1995.

Schumann, U. "In Situ Observations of Particles in Jet Aircraft Exhausts and Contrails for Different Sulfur-Containing Fuels." *Journal of Geographical Research*, Vol. 101, Issue D3, 1996.

Scientists for Global Responsibility. "Climate Engineering: A Critical Review of Proposals." School of Environmental Sciences. UEA, Norwich, November 1996.

Shachtman, Noah. "Pentagon Preps for War in Space." *WiredNews. com*, February 2004.

Shapley, Deborah. "Weather Warfare: Pentagon Concedes 7-Year Vietnam Effort." *Science*, June 7, 1974.

Simpson, R.H. "Structure of an Immature Hurricane" *Bull. of Amer. Meteo.Soc.* vol.35 no.8 pp.335-350, 1954.

Simpson, R. H., Joanne S. Malkus. "Experiments in Hurricane Modification," *Scientific American*, December 1964.

Snel, Alan. "Sailors - 'We Were Used' - Vets Exposed To Toxic Agents Want Answers, Justice." Originally published in *Florida Today*, reprinted in *The Federal Observer*, Vol. 06, No. 105, April 16, 2006.

Solomatin, Yuru; Translated by Dmitry Sudakov. "HAARP Poses Global Threat." *Pravda*, January 15, 2003.

Spokesman Review. "After 12-year Wait for Trial, Downwinders Losing Hope," May 18, 2003.

Spokesman Review. "Hanford Plaintiffs Seek Details," April 2, 2004.

Spokesman Review. "Hanford Put Area At Risk: Spokane, North Idaho Were Exposed to Significant Radiation." April 22, 1994.

Sutherland, Robert A. "Results of Man-Made Fog Experiment," *Proceedings of the 1991 Battlefield Atmospherics Conference*, Fort Bliss, Tex.: Hinman Hall, 3-6 December 1991.

Standler, Ronald B. "History and Problems in Weather Modification" *http://www.rbs2.com/w2.htm* revised 21 Jan 2003.

Standler, Ronald B. "Weather Modification Law in the USA" *http://www.rbs2.com/weather.htm* revised 23 Dec 2002.

Stipp, David. "Climate Collapse, The Pentagon's Weather Nightmare Could Change Radically and Fast," *Fortune Magazine*, February 9, 2004.

Swartz, Tim R. "Meteorological Madness: Is Weather Being Used As The Ultimate Weapon?" *Mysteries* Vol. 4 No. 1, Feb-April 2006.

Teller, Edward. "The Planet Needs a Sunscreen" *Wall Street Journal*, October 17, 1997.

Tarasofsky R. "Legal Protection of the Environment during International Armed Conflicts," in *The Netherlands Yearbook of*

International Law, v. XXIV, p. 51, 1993.

Tomlinson, Edward M., Kenneth C. Young, and Duane D. Smith "Laser Technology Applications for Dissipation of Warm Fog at Airfields, PL-TR-92-2087." Hanscom AFB, Mass.: Air Force Materiel Command, 1992.

Townsend, Mark, Paul Harris. "Pentagon Tells Bush: Climate Change Will Destroy Us," *The Observer*, February 24, 2004.

Travis, D.J. "Diurnal Temperature Range Modifications Induced By Contrails." *Proc. of the 13th Conf. Planned and Inadvertent Wea. Mod.*, Atlanta, GA. 1996

Travis, D.J., A.M. Carleton, and S.A. Changnon. "An Empirical Model To Predict Widespread Occurrences Of Contrails." *J. Appl. Meteor.*, 36, 1997.

Trustees for Alaska. "Waking up the Military." *Earthpulse Flashpoints*, Series 1 Volume 2.

Twietmeyer, Ted. "Chemtrails And Aluminum Harm Heart, Lungs And Plants." *Rense.com,* January 2006.

Theroux, Michael. "Space Based Weather Control: The 'Thunderstorm Solar Power Satellite.'" www.borderlands.com/spacewea.htm.

Thomas, William. "Stolen Skies: The Chemtrail Mystery." *Earth Island Journal* Summer 2002 Vol. 17, No. 2.

USAToday.com "Asian dust brings hazy skies to western U.S." April 18, 2001.

Utlaut, W.F. and R. Cohen. "Modifying the Ionosphere with Intense Radio Waves." *Science*, 1971.

Vonnegut, Bernard. "The Nucleation of Ice Formation by Silver Iodide," *J. Applied Physics,* Vol. 18, pp. 593-595, July 1947.

Vonnegut, Bernard. "Nucleation of Supercooled Water Clouds

373

by Silver Iodide Smokes," *Chemical Reviews,* Vol. 44, pp. 277-289, April 1949.

Vonnegut, Bernard and Kiah Maynard, "Spray-Nozzle Type Silver-Iodide Smoke Generator for Airplane Use," *Bulletin of the American Meteorological Society,* Vol. 33, pp. 420-428, December 1952.

Watson, Traci. "Plane Trails in Sky Turn Up the Heat Below, Study Suggests." *USA Today.* April 29, 2004.

Wall Street Journal, "Raytheon to Acquire E-Systems for $64 a Share...", Pages A1, A3, and A16, April 3, 1995.

Watts, Jonathan. "China plans to double air traffic with 100 new aircraft a year." *The Guardian*, February 15, 2006.

Whiteford, Gary: "Nuclear Bomb Tests and Earthquakes: Dangerous Patterns and Trends." *Pulse of the Planet* 2:10-21 Fall 1989.

Willoughby, H. E., D. P. Jorgensen, R. A. Black, and S. L. Rosenthal. "Project STORMFURY, A Scientific Chronicle, 1962-1983." *Bull. Amer. Meteor. Soc.*, 66, 505-514, 1985.

Wong and Brandt. "Ionospheric Modification - An Outdoor Laboratory For Plasma And Atmospheric Science." *Radio Science*, 25(6), 1990.

Worthington, Amy. "Chemtrails and Terror in the Age of Nuclear War." *Idaho Observer*, May 2004.

Worthington, Amy. "Evidence: Chemtrails Include Hazardous Barium Compounds." *Idaho Observer*, November 2000.

Worthington, Amy. "Reading between the lines in the sky." *Idaho Observer*, February 2001.

Worthington, Amy. "Strange Haze and Smart Alecs." *Idaho Observer.* June 2000.

Yuzon, F. "Deliberate Environmental Modification Through the Use of Chemical and Biological Weapons: 'Greening' the International Laws of Armed Conflict to Establish an Environmentally Protective

374

Regime." *Am. U. J. Inter'l. L & Policy*, v. 11:5, p. 804, 1996.

WEBSITES OF INTEREST

CONTRAIL/CHEMTRAIL

Aerosol Operation Crimes & Cover-Up
http://www.carnicom.com/contrails.htm

Barium Blues
http://www.bariumblues.com/

CACTUS (Citizens Against Chemtrails U.S.)
http://www.geocities.com/cactusmailbox/CACTUS.html

Canadian Chemtrails
http://www.geocities.com/canadianchemtrails/

Chemtrail Action Network
http://www.chemtrailactionnetwork.com/

Chemtrail Campaigners
http://members.lycos.co.uk/fotdragon/articles/outbreaks.html

Chemtrail Central
http://www.chemtrailcentral.com/

Chemtrail Resources
http://www.mysteriousnewzealand.co.nz/resources/chemtrails.html

Chemtrails 911
http://www.skyhighway.com/~chemtrails/

Chemtrails DataPage
http://www.rense.com/politics6/chemdatapage.html

Chemtrails Hall Of Shame
http://www.lightwatcher.com/chemtrails/hos.html

Weather Warfare

Chemtrails Over America
http://home1.gte.net/quakker/documents/chemtrails_over_
america.htm#coa

Cloud-Busters.com
http://archive.cloud-busters.com/

Contrails, or 'Trails-CON?
http://goodsky.homestead.com/files/index.html

Educate-Yourself
http://educate-yourself.org

The Holmestead
http://www.holmestead.ca/

New Mexicans for Science and Reason
http://www.nmsr.org/chemtrls.htm

Strange Days Strange Skies
http://imageevent.com/firesat/strangedaysstrangeskies

Strange Haze
http://strangehaze.freeservers.com/index.html

Tracers
http://tracers.8m.com/

HAARP

Brother Jonathan Gazette
http://www.brojon.org/frontpage.html

Columbia's Sacrifice
http://www.columbiassacrifice.com/

Defense Advanced Research Projects Agency (DARPA)
http://www.darpa.mil/

Earthpulse Press
http://www.earthpulse.com/

Eastlund Scientific Enterprises Corporation (ESEC)
http://www.eastlundscience.com/

H.A.A.R.P.
http://www.alaska.net/~logjam/HAARP.html

HAARP
http://www.crystalinks.com/haarp.html

HAARP.NET
http://www.haarp.net/

HIGH FREQUENCY ACTIVE AURORAL RESEARCH PROGRAM
http://www.haarp.alaska.edu/haarp/

High Frequency Active Auroral Research Project (HAARP)
http://www.darpa.mil/tto/programs/haarp.htm

Holes in Heaven? A Documentary
http://www.haarp.com/

RadarMatrix
http://www.radarmatrix.com/

SRI Int'l Ionospheric and Space Sciences Group
http://isr.sri.com/iono/issmain.html

Tactical Technology Office of DARPA
http://www.darpa.mil/tto/

DEFENCE AND SPACE NEWS

Defence News.com
http://www.defencenews.com/

Defence Systems Daily
http://defence-data.com/index2/index2.shtml

Jane's Defence Weekly
http://jdw.janes.com/public/jdw/index.shtml

Jane's Information Group
http://www.janes.com/

National Defense Magazine
http://www.nationaldefensemagazine.org/

SatNews Weekly Online
http://www.satnews.com/main.htm

Space.com
http://www.space.com/

SpaceDaily
http://www.spacedaily.com/

Space Environment Center
http://www.sec.noaa.gov/

SpaceWeather.com
http://www.spaceweather.com/

StarDate Online
http://stardate.org/

WEATHER CONTROL & ENVIRONMENTAL WARFARE

Air Force 2025
http://csat.au.af.mil/2025/index.htm

Atmospheric and Environmental Research, Inc.
http://www.aer.com/home/home.html

Climate Engineering
http://www.chooseclimate.org/cleng/

Climate TimeLine
http://www.ngdc.noaa.gov/paleo/ctl/index.html

Depleted Uranium Watch
http://www.stopnato.org.uk/du-watch/index.htm

EcoNews: Environmental War Desk
http://www.ecologynews.com/cuenewsdesk.html

Edmonds Institute
http://www.edmonds-institute.org/index.html

ELFRAD Group
http://www.elfrad.com/

Flash Radar
http://members.toast.net/flashradar/flashx.html

Ice Crystal Engineering, LLC
http://www.iceflares.com/

Liberty Exposure: WEATHER WEAPONS
http://www.alphalink.com.au/~noelmcd/weather.htm

North American Weather Consultants
http://www.nawcinc.com/index.html

The Tom Bearden Website
http://www.cheniere.org/

The Weather Master
http://twm.co.nz/index.html

Weather Modification, Inc.
http://www.weathermod.com/

Weather Wars
http://www.weatherwars.info/index.php

OTHER

Agenda 21
http://www.un.org/esa/sustdev/documents/agenda21/index.htm

American Meteorological Society
http://www.ametsoc.org/

379

Atmospheric Chemistry Program
http://www.atmos.anl.gov/ACP/

Atmospheric Radiation Measurement
http://www.arm.gov/

Atmospheric Science Program
http://www.asp.bnl.gov/

Carbon Dioxide Information Analysis Center
http://cdiac.esd.ornl.gov/

Center for Aerosol and Cloud Chemistry
http://www.aerodyne.com/cacc/cacc.html

Center for Atmospheric and Environmental Chemistry
http://www.aerodyne.com/caec/caec.html

The Earth Charter
http://www.earthcharter.org/

EMFacts Consultancy
http://www.emfacts.com/

The Federation of American Scientists
http://www.fas.org/main/home.jsp

Freedom 21
http://www.freedom21.org/

Freedom 21 Santa Cruz
http://www.freedom21santacruz.net/site/

Global Network Against Weapons and Nuclear Power in Space
http://www.space4peace.org/

Global Warming International Center
http://www.globalwarming.net/

Green Cross International

http://gcinwa.newaccess.ch/indexmacie.htm

Institute of Global Environment and Society
http://grads.iges.org/home.html

Intergovernmental Panel on Climate Change
http://www.ipcc.ch/

Military Toxics Project
http://www.miltoxproj.org/

Missile Defense Agency
http://www.mda.mil/mdalink/html/mdalink.html

NARSTO
http://www.narsto.org/

National Academy of Sciences, Board on Atmospheric Sciences and Climate
http://dels.nas.edu/basc/

National Oceanic & Atmospheric Administration (NOAA)
http://www.noaa.gov/

The Ocean and Climate Change Institute
http://www.whoi.edu/institutes/occi/viewTopic.
do?o=read&id=501

Paleoclimatology Branch of the National Climatic Data Center
http://www.ncdc.noaa.gov/paleo/index.html

Paragon Foundation
http://www.paragonfoundation.org/

Program for Climate Model Diagnosis and Intercomparison
http://www-pcmdi.llnl.gov/

Radiation and Public Health Project
http://www.radiation.org/

The RMA Debate
http://www.comw.org/rma/

Stewards of the Range
http://www.stewardsoftherange.org/

The Sunshine Project
http://www.sunshine-project.org/

Tropospheric Aerosol Program
http://www.asp.bnl.gov/

Tyndall Centre for Climate Change Research
http://www.tyndall.ac.uk/index.shtml

U.S. Global Change Research Program
http://globalchange.gov/

Union of Concerned Scientists
http://www.ucsusa.org/

United Nations Framework Convention On Climate Change
http://unfccc.int/2860.php

The Wildlands Project
http://www.twp.org:80/cms/index.cfm?group_id=1000

Wildlands Project Revealed: Citizens With Common Sense
http://www.wildlandsprojectrevealed.org/

The Woods Hole Research Center
http://www.whrc.org/

World Environment Center
http://www.wec.org/

World Resources Institute
http://www.wri.org/index.html

Worldwatch Institute
http://www.worldwatch.org/

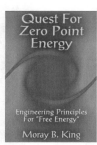

QUEST FOR ZERO-POINT ENERGY
Engineering Principles for "Free Energy"
by Moray B. King

King expands, with diagrams, on how free energy and anti-gravity are possible. The theories of zero point energy maintain there are tremendous fluctuations of electrical field energy embedded within the fabric of space. King explains the following topics: Tapping the Zero-Point Energy as an Energy Source; Fundamentals of a Zero-Point Energy Technology; Vacuum Energy Vortices; The Super Tube; Charge Clusters: The Basis of Zero-Point Energy Inventions; Vortex Filaments, Torsion Fields and the Zero-Point Energy; Transforming the Planet with a Zero-Point Energy Experiment; Dual Vortex Forms: The Key to a Large Zero-Point Energy Coherence. Packed with diagrams, patents and photos. With power shortages now a daily reality in many parts of the world, this book offers a fresh approach very rarely mentioned in the mainstream media.
224 PAGES. 6X9 PAPERBACK. ILLUSTRATED. $14.95. CODE: QZPE

TAPPING THE ZERO POINT ENERGY
Free Energy & Anti-Gravity in Today's Physics
by Moray B. King

King explains how free energy and anti-gravity are possible. The theories of the zero point energy maintain there are tremendous fluctuations of electrical field energy imbedded within the fabric of space. This book tells how, in the 1930s, inventor T. Henry Moray could produce a fifty kilowatt "free energy" machine; how an electrified plasma vortex creates anti-gravity; how the Pons/Fleischmann "cold fusion" experiment could produce tremendous heat without fusion; and how certain experiments might produce a gravitational anomaly.
180 PAGES. 5X8 PAPERBACK. ILLUSTRATED. $12.95. CODE: TAP

THE FREE-ENERGY DEVICE HANDBOOK
A Compilation of Patents and Reports
by David Hatcher Childress

A large-format compilation of various patents, papers, descriptions and diagrams concerning free-energy devices and systems. *The Free-Energy Device Handbook* is a visual tool for experimenters and researchers into magnetic motors and other "over-unity" devices. With chapters on the Adams Motor, the Hans Coler Generator, cold fusion, superconductors, "N" machines, space-energy generators, Nikola Tesla, T. Townsend Brown, and the latest in free-energy devices. Packed with photos, technical diagrams, patents and fascinating information, this book belongs on every science shelf. With energy and profit being a major political reason for fighting various wars, free-energy devices, if ever allowed to be mass distributed to consumers, could change the world! Get your copy now before the Department of Energy bans this book!
292 PAGES. 8X10 PAPERBACK. ILLUSTRATED. BIBLIOGRAPHY. $16.95. CODE: FEH

ETHER TECHNOLOGY
A Rational Approach to Gravity Control
by Rho Sigma

This classic book on anti-gravity and free energy is back in print and back in stock. Written by a well-known American scientist under the pseudonym of "Rho Sigma," this book delves into international efforts at gravity control and discoid craft propulsion. Before the Quantum Field, there was "Ether." This small, but informative book has chapters on John Searle and "Searle discs;" T. Townsend Brown and his work on anti-gravity and ether-vortex turbines. Includes a forward by former NASA astronaut Edgar Mitchell.
108 PAGES. 6X9 PAPERBACK. ILLUSTRATED. $12.95. CODE: ETT

THE TIME TRAVEL HANDBOOK
A Manual of Practical Teleportation & Time Travel
edited by David Hatcher Childress

In the tradition of *The Anti-Gravity Handbook* and *The Free-Energy Device Handbook*, science and UFO author David Hatcher Childress takes us into the weird world of time travel and teleportation. Not just a whacked-out look at science fiction, this book is an authoritative chronicling of real-life time travel experiments, teleportation devices and more. *The Time Travel Handbook* takes the reader beyond the government experiments and deep into the uncharted territory of early time travellers such as Nikola Tesla and Guglielmo Marconi and their alleged time travel experiments, as well as the Wilson Brothers of EMI and their connection to the Philadelphia Experiment—the U.S. Navy's forays into invisibility, time travel, and teleportation. Childress looks into the claims of time travelling individuals, and investigates the unusual claim that the pyramids on Mars were built in the future and sent back in time. A highly visual, large format book, with patents, photos and schematics. Be the first on your block to build your own time travel device!
316 PAGES. 7X10 PAPERBACK. ILLUSTRATED. $16.95. CODE: TTH

MAN-MADE UFOS 1944—1994
Fifty Years of Suppression
by Renato Vesco & David Hatcher Childress

A comprehensive look at the early "flying saucer" technology of Nazi Germany and the genesis of man-made UFOs. This book takes us from the work of captured German scientists to escaped battalions of Germans, secret communities in South America and Antarctica to todays state-of-the-art "Dreamland" flying machines. Heavily illustrated, this astonishing book blows the lid off the "government UFO conspiracy" and explains with technical diagrams the technology involved. Examined in detail are secret underground airfields and factories; German secret weapons; "suction" aircraft; the origin of NASA; gyroscopic stabilizers and engines; the secret Marconi aircraft factory in South America; and more. Introduction by W.A. Harbinson, author of the Dell novels *GENESIS* and *REVELATION*.
318 PAGES. 6X9 PAPERBACK. ILLUSTRATED. INDEX & FOOTNOTES. $18.95. CODE: MMU

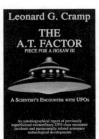

THE A.T. FACTOR
A Scientists Encounter with UFOs: Piece For A Jigsaw Part 3
by Leonard Cramp

British aerospace engineer Cramp began much of the scientific anti-gravity and UFO propulsion analysis back in 1955 with his landmark book *Space, Gravity & the Flying Saucer* (out-of-print and rare). His next books (available from Adventures Unlimited) *UFOs & Anti-Gravity: Piece for a Jig-Saw* and *The Cosmic Matrix: Piece for a Jig-Saw Part 2* began Cramp's in depth look into gravity control, free-energy, and the interlocking web of energy that pervades the universe. In this final book, Cramp brings to a close his detailed and controversial study of UFOs and Anti-Gravity.
324 PAGES. 6X9 PAPERBACK. ILLUSTRATED. BIBLIOGRAPHY. INDEX. $16.95. CODE: ATF

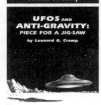

COSMIC MATRIX
Piece for a Jig-Saw, Part Two
by Leonard G. Cramp

Leonard G. Cramp, a British aerospace engineer, wrote his first book *Space Gravity and the Flying Saucer* in 1954. Cosmic Matrix is the long-awaited sequel to his 1966 book *UFOs & Anti-Gravity: Piece for a Jig-Saw*. Cramp has had a long history of examining UFO phenomena and has concluded that UFOs use the highest possible aeronautic science to move in the way they do. Cramp examines anti-gravity effects and theorizes that this super-science used by the craft—described in detail in the book—can lift mankind into a new level of technology, transportation and understanding of the universe. The book takes a close look at gravity control, time travel, and the interlocking web of energy between all planets in our solar system with Leonard's unique technical diagrams. A fantastic voyage into the present and future!
364 PAGES. 6X9 PAPERBACK. ILLUSTRATED. BIBLIOGRAPHY. $16.00. CODE: CMX

UFOS AND ANTI-GRAVITY
Piece For A Jig-Saw
by Leonard G. Cramp

Leonard G. Cramp's 1966 classic book on flying saucer propulsion and suppressed technology is a highly technical look at the UFO phenomena by a trained scientist. Cramp first introduces the idea of 'anti-gravity' and introduces us to the various theories of gravitation. He then examines the technology necessary to build a flying saucer and examines in great detail the technical aspects of such a craft. Cramp's book is a wealth of material and diagrams on flying saucers, anti-gravity, suppressed technology, G-fields and UFOs. Chapters include Crossroads of Aerodynamics, Aerodynamic Saucers, Limitations of Rocketry, Gravitation and the Ether, Gravitational Spaceships, G-Field Lift Effects, The Bi-Field Theory, VTOL and Hovercraft, Analysis of UFO photos, more.
388 PAGES. 6X9 PAPERBACK. ILLUSTRATED. $16.95. CODE: UAG

THE TESLA PAPERS
Nikola Tesla on Free Energy & Wireless Transmission of Power
by Nikola Tesla, edited by David Hatcher Childress

David Hatcher Childress takes us into the incredible world of Nikola Tesla and his amazing inventions. Tesla's rare article "The Problem of Increasing Human Energy with Special Reference to the Harnessing of the Sun's Energy" is included. This lengthy article was originally published in the June 1900 issue of *The Century Illustrated Monthly Magazine* and it was the outline for Tesla's master blueprint for the world. Tesla's fantastic vision of the future, including wireless power, anti-gravity, free energy and highly advanced solar power. Also included are some of the papers, patents and material collected on Tesla at the Colorado Springs Tesla Symposiums, including papers on: •The Secret History of Wireless Transmission •Tesla and the Magnifying Transmitter •Design and Construction of a Half-Wave Tesla Coil •Electrostatics: A Key to Free Energy •Progress in Zero-Point Energy Research •Electromagnetic Energy from Antennas to Atoms •Tesla's Particle Beam Technology •Fundamental Excitatory Modes of the Earth-Ionosphere Cavity
325 PAGES. 8X10 PAPERBACK. ILLUSTRATED. $16.95. CODE: TTP

THE FANTASTIC INVENTIONS OF NIKOLA TESLA
by Nikola Tesla with additional material by David Hatcher Childress

This book is a readable compendium of patents, diagrams, photos and explanations of the many incredible inventions of the originator of the modern era of electrification. In Tesla's own words are such topics as wireless transmission of power, death rays, and radio-controlled airships. In addition, rare material on German bases in Antarctica and South America, and a secret city built at a remote jungle site in South America by one of Tesla's students, Guglielmo Marconi. Marconi's secret group claims to have built flying saucers in the 1940s and to have gone to Mars in the early 1950s! Incredible photos of these Tesla craft are included. The Ancient Atlantean system of broadcasting energy through a grid system of obelisks and pyramids is discussed, and a fascinating concept comes out of one chapter: that Egyptian engineers had to wear protective metal head-shields while in these power plants, hence the Egyptian Pharoah's head covering as well as the Face on Mars! •His plan to transmit free electricity into the atmosphere. •How electrical devices would work using only small antennas. •Why unlimited power could be utilized anywhere on earth. •How radio and radar technology can be used as death-ray weapons in Star Wars.
342 PAGES. 6x9 PAPERBACK. ILLUSTRATED. $16.95. CODE: FINT

REICH OF THE BLACK SUN
Nazi Secret Weapons and the Cold War Allied Legend
by Joseph P. Farrell

Why were the Allies worried about an atom bomb attack by the Germans in 1944? Why did the Soviets threaten to use poison gas against the Germans? Why did Hitler in 1945 insist that holding Prague could win the war for the Third Reich? Why did US General George Patton's Third Army race for the Skoda works at Pilsen in Czechoslovakia instead of Berlin? Why did the US Army not test the uranium atom bomb it dropped on Hiroshima? Why did the Luftwaffe fly a non-stop round trip mission to within twenty miles of New York City in 1944? *Reich of the Black Sun* takes the reader on a scientific-historical journey in order to answer these questions. Arguing that Nazi Germany actually won the race for the atom bomb in late 1944, and then goes on to explore the even more secretive research the Nazis were conducting into the occult, alternative physics and new energy sources.
352 PAGES. 6x9 PAPERBACK. ILLUSTRATED. BIBLIOGRAPHY. $16.95. CODE: ROBS

SAUCERS OF THE ILLUMINATI
by Jim Keith, Foreword by Kenn Thomas

Seeking the truth behind stories of alien invasion, secret underground bases, and the secret plans of the New World Order, *Saucers of the Illuminati* offers ground breaking research, uncovering clues to the nature of UFOs and to forces even more sinister: the secret cabal behind planetary control! Includes mind control, saucer abductions, the MJ-12 documents, cattle mutilations, government anti-gravity testing, the Sirius Connection, science fiction author Philip K. Dick and his efforts to expose the Illuminati, plus more from veteran conspiracy and UFO author Keith. Conspiracy expert Keith's final book on UFOs and the highly secret group that manufactures them and uses them for their own purposes: the control and manipulation of the population of planet Earth.
148 PAGES. 6x9 PAPERBACK. ILLUSTRATED. $12.95. CODE: SOIL

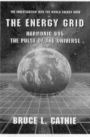

THE ENERGY MACHINE OF T. HENRY MORAY
by Moray B. King

In the 1920s T. Henry Moray invented a "free energy" device that reportedly output 50 kilowatts of electricity. It could not be explained by standard science at that time. The electricity exhibited a strange "cold current" characteristic where thin wires could conduct appreciable power without heating. Moray suffered ruthless suppression, and in 1939 the device was destroyed. Frontier science lecturer and author Moray B. King explains the invention with today's science. Modern physics recognizes that the vacuum contains tremendous energy called the zero-point energy. A way to coherently activate it appears surprisingly simple: first create a glow plasma or corona, then abruptly pulse it. Other inventors have discovered this approach (sometimes unwittingly) to create novel energy devices, and they too were suppressed. The common pattern of their technologies clarified the fundamental operating principle. King hopes to inspire engineers and inventors so that a new energy source can become available to mankind.
192 PAGES. 6x8 PAPERBACK. ILLUSTRATED. $14.95. CODE: EMHM

THE ENERGY GRID
Harmonic 695, The Pulse of the Universe
by Captain Bruce Cathie.

This is the breakthrough book that explores the incredible potential of the Energy Grid and the Earth's Unified Field all around us. Cathie's first book, *Harmonic 33*, was published in 1968 when he was a commercial pilot in New Zealand. Since then, Captain Bruce Cathie has been the premier investigator into the amazing potential of the infinite energy that surrounds our planet every microsecond. Cathie investigates the Harmonics of Light and how the Energy Grid is created. In this amazing book are chapters on UFO Propulsion, Nikola Tesla, Unified Equations, the Mysterious Aerials, Pythagoras & the Grid, Nuclear Detonation and the Grid, Maps of the Ancients, an Australian Stonehenge examined, more.
255 PAGES. 6x9 TRADEPAPER. ILLUSTRATED. $15.95. CODE: TEG

THE BRIDGE TO INFINITY
Harmonic 371244
by Captain Bruce Cathie

Cathie has popularized the concept that the earth is crisscrossed by an electromagnetic grid system that can be used for anti-gravity, free energy, levitation and more. The book includes a new analysis of the harmonic nature of reality, acoustic levitation, pyramid power, harmonic receiver towers and UFO propulsion. It concludes that today's scientists have at their command a fantastic store of knowledge with which to advance the welfare of the human race.
204 PAGES. 6x9 TRADEPAPER. ILLUSTRATED. $14.95. CODE: BTF

THE HARMONIC CONQUEST OF SPACE
by Captain Bruce Cathie

Chapters include: Mathematics of the World Grid; the Harmonics of Hiroshima and Nagasaki; Harmonic Transmission and Receiving; the Link Between Human Brain Waves; the Cavity Resonance between the Earth; the Ionosphere and Gravity; Edgar Cayce—the Harmonics of the Subconscious; Stonehenge; the Harmonics of the Moon; the Pyramids of Mars; Nikola Tesla's Electric Car; the Robert Adams Pulsed Electric Motor Generator; Harmonic Clues to the Unified Field; and more. Also included are tables showing the harmonic relations between the earth's magnetic field, the speed of light, and anti-gravity/gravity acceleration at different points on the earth's surface. New chapters in this edition on the giant stone spheres of Costa Rica, Atomic Tests and Volcanic Activity, and a chapter on Ayers Rock analysed with Stone Mountain, Georgia.
248 PAGES. 6x9. PAPERBACK. ILLUSTRATED. BIBLIOGRAPHY. $16.95. CODE: HCS

THE ANTI-GRAVITY HANDBOOK
edited by David Hatcher Childress, with Nikola Tesla, T.B. Paulicki, Bruce Cathie, Albert Einstein and others

The new expanded compilation of material on Anti-Gravity, Free Energy, Flying Saucer Propulsion, UFOs, Suppressed Technology, NASA Cover-ups and more. Highly illustrated with patents, technical illustrations and photos. This revised and expanded edition has more material, including photos of Area 51, Nevada, the government's secret testing facility. This classic on weird science is back in a 90s format!
- **How to build a flying saucer.**
- **Arthur C. Clarke on Anti-Gravity.**
- **Crystals and their role in levitation.**
- **Secret government research and development.**
- **Nikola Tesla on how anti-gravity airships could draw power from the atmosphere.**
- **Bruce Cathie's Anti-Gravity Equation.**
- **NASA, the Moon and Anti-Gravity.**

230 PAGES. 7x10 PAPERBACK. ILLUSTRATED. $14.95. CODE: **AGH**

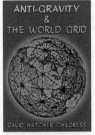

ANTI–GRAVITY & THE WORLD GRID

Is the earth surrounded by an intricate electromagnetic grid network offering free energy? This compilation of material on ley lines and world power points contains chapters on the geography, mathematics, and light harmonics of the earth grid. Learn the purpose of ley lines and ancient megalithic structures located on the grid. Discover how the grid made the Philadelphia Experiment possible. Explore the Coral Castle and many other mysteries, including acoustic levitation, Tesla Shields and scalar wave weaponry. Browse through the section on anti-gravity patents, and research resources.

274 PAGES. 7x10 PAPERBACK. ILLUSTRATED. $14.95. CODE: **AGW**

ANTI–GRAVITY & THE UNIFIED FIELD
edited by David Hatcher Childress

Is Einstein's Unified Field Theory the answer to all of our energy problems? Explored in this compilation of material is how gravity, electricity and magnetism manifest from a unified field around us. Why artificial gravity is possible; secrets of UFO propulsion; free energy; Nikola Tesla and anti-gravity airships of the 20s and 30s; flying saucers as superconducting whirls of plasma; anti-mass generators; vortex propulsion; suppressed technology; government cover-ups; gravitational pulse drive; spacecraft & more.

240 PAGES. 7x10 PAPERBACK. ILLUSTRATED. $14.95. CODE: **AGU**

THE GIZA DEATH STAR
The Paleophysics of the Great Pyramid & the Military Complex at Giza
by Joseph P. Farrell

Physicist Joseph Farrell's amazing book on the secrets of Great Pyramid of Giza. *The Giza Death Star* starts where British engineer Christopher Dunn leaves off in his 1998 book, *The Giza Power Plant*. Was the Giza complex part of a military installation over 10,000 years ago? Chapters include: An Archaeology of Mass Destruction, Thoth and Theories; The Machine Hypothesis; Pythagoras, Plato, Planck, and the Pyramid; The Weapon Hypothesis; Encoded Harmonics of the Planck Units in the Great Pyramid; High Freguency Direct Current "Impulse" Technology; The Grand Gallery and its Crystals: Gravito-acoustic Resonators; The Other Two Large Pyramids; the "Causeways," and the "Temples"; A Phase Conjugate Howitzer; Evidence of the Use of Weapons of Mass Destruction in Ancient Times; more.

290 PAGES. 6x9 PAPERBACK. ILLUSTRATED. $16.95. CODE: **GDS**

DARK MOON
Apollo and the Whistleblowers
by Mary Bennett and David Percy

- Was Neil Armstrong really the first man on the Moon?
- Did you know a second craft was going to the Moon at the same time as Apollo 11?
- Do you know that potentially lethal radiation is prevalent throughout deep space?
- Do you know there are serious discrepancies in the account of the Apollo 13 'accident'?
- Did you know that 'live' color TV from the Moon was not actually live at all?
- Did you know that the Lunar Surface Camera had no viewfinder?
- Do you know that lighting was used in the Apollo photographs—yet no lighting equipment was taken to the Moon?

All these questions, and more, are discussed in great detail by British researchers Bennett and Percy in *Dark Moon*, the definitive book (nearly 600 pages) on the possible faking of the Apollo Moon missions. Bennett and Percy delve into every possible aspect of this beguiling theory, one that rocks the very foundation of our beliefs concerning NASA and the space program. Tons of NASA photos analyzed for possible deceptions.

568 PAGES. 6x9 PAPERBACK. ILLUSTRATED. BIBLIOGRAPHY. INDEX. $25.00. CODE: **DMO**

TECHNOLOGY OF THE GODS
The Incredible Sciences of the Ancients
by David Hatcher Childress

Popular *Lost Cities* author David Hatcher Childress takes us into the amazing world of ancient technology, from computers in antiquity to the "flying machines of the gods." Childress looks at the technology that was allegedly used in Atlantis and the theory that the Great Pyramid of Egypt was originally a gigantic power station. He examines tales of ancient flight and the technology that it involved; how the ancients used electricity; megalithic building techniques; the use of crystal lenses and the fire from the gods; evidence of various high tech weapons in the past, including atomic weapons; ancient metallurgy and heavy machinery; the role of modern inventors such as Nikola Tesla in bringing ancient technology back into modern use; impossible artifacts; and more.

356 PAGES. 6X9 PAPERBACK. ILLUSTRATED. BIBLIOGRAPHY. $16.95. CODE: TGOD

VIMANA AIRCRAFT OF ANCIENT INDIA & ATLANTIS
by David Hatcher Childress, introduction by Ivan T. Sanderson

Did the ancients have the technology of flight? In this incredible volume on ancient India, authentic Indian texts such as the *Ramayana* and the *Mahabharata* are used to prove that ancient aircraft were in use more than four thousand years ago. Included in this book is the entire Fourth Century BC manuscript *Vimaanika Shastra* by the ancient author Maharishi Bharadwaaja, translated into English by the Mysore Sanskrit professor G.R. Josyer. Also included are chapters on Atlantean technology, the incredible Rama Empire of India and the devastating wars that destroyed it. Also an entire chapter on mercury vortex propulsion and mercury gyros, the power source described in the ancient Indian texts. Not to be missed by those interested in ancient civilizations or the UFO enigma.

334 PAGES. 6X9 PAPERBACK. ILLUSTRATED. $15.95. CODE: VAA

LOST CONTINENTS & THE HOLLOW EARTH
I Remember Lemuria and the Shaver Mystery
by David Hatcher Childress & Richard Shaver

Lost Continents & the Hollow Earth is Childress' thorough examination of the early hollow earth stories of Richard Shaver and the fascination that fringe fantasy subjects such as lost continents and the hollow earth have had for the American public. Shaver's rare 1948 book *I Remember Lemuria* is reprinted in its entirety, and the book is packed with illustrations from Ray Palmer's *Amazing Stories* magazine of the 1940s. Palmer and Shaver told of tunnels running through the earth—tunnels inhabited by the Deros and Teros, humanoids from an ancient spacefaring race that had inhabited the earth, eventually going underground, hundreds of thousands of years ago. Childress discusses the famous hollow earth books and delves deep into whatever reality may be behind the stories of tunnels in the earth. Operation High Jump to Antarctica in 1947 and Admiral Byrd's bizarre statements, tunnel systems in South America and Tibet, the underground world of Agartha, the belief of UFOs coming from the South Pole, more.

344 PAGES. 6X9 PAPERBACK. ILLUSTRATED. $16.95. CODE: LCHE

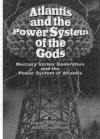

ATLANTIS & THE POWER SYSTEM OF THE GODS
Mercury Vortex Generators & the Power System of Atlantis
by David Hatcher Childress and Bill Clendenon

Atlantis and the Power System of the Gods starts with a reprinting of the rare 1990 book *Mercury: UFO Messenger of the Gods* by Bill Clendenon. Clendenon takes on an unusual voyage into the world of ancient flying vehicles, strange personal UFO sightings, a meeting with a "Man In Black" and then to a centuries-old library in India where he got his ideas for the diagrams of mercury vortex engines. The second part of the book is Childress' fascinating analysis of Nikola Tesla's broadcast system in light of Edgar Cayce's "Terrible Crystal" and the obelisks of ancient Egypt and Ethiopia. Includes: Atlantis and its crystal power towers that broadcast energy; how these incredible power stations may still exist today; inventor Nikola Tesla's nearly identical system of power transmission; Mercury Proton Gyros and mercury vortex propulsion; more. Richly illustrated, and packed with evidence that Atlantis not only existed—it had a world-wide energy system more sophisticated than ours today.

246 PAGES. 6X9 PAPERBACK. ILLUSTRATED. $15.95. CODE: APSG

A HITCHHIKER'S GUIDE TO ARMAGEDDON
by David Hatcher Childress

With wit and humor, popular Lost Cities author David Hatcher Childress takes us around the world and back in his trippy finalé to the Lost Cities series. He's off on an adventure in search of the apocalypse and end times. Childress hits the road from the fortress of Megiddo, the legendary citadel in northern Israel where Armageddon is prophesied to start. Hitchhiking around the world, Childress takes us from one adventure to another, to ancient cities in the deserts and the legends of worlds before our own. Childress muses on the rise and fall of civilizations, and the forces that have shaped mankind over the millennia, including wars, invasions and cataclysms. He discusses the ancient Armageddons of the past, and chronicles recent Middle East developments and their ominous undertones. In the meantime, he becomes a cargo cult god on a remote island off New Guinea, gets dragged into the Kennedy Assassination by one of the "conspirators," investigates a strange power operating out of the Altai Mountains of Mongolia, and discovers how the Knights Templar and their off-shoots have driven the world toward an epic battle centered around Jerusalem and the Middle East.

320 PAGES. 6X9 PAPERBACK. ILLUSTRATED. BIBLIOGRAPHY. INDEX. $16.95. CODE: HGA

One Adventure Place
P.O. Box 74
Kempton, Illinois 60946
United States of America
Tel.: 815-253-6390 • Fax: 815-253-6300
Email: auphq@frontiernet.net
http://www.adventuresunlimitedpress.com
or www.adventuresunlimited.nl

ORDERING INSTRUCTIONS

➤➤ Remit by USD$ Check, Money Order or Credit Card

➤➤ Visa, Master Card, Discover & AmEx Accepted

➤➤ Prices May Change Without Notice

➤➤ 10% Discount for 3 or more Items

SHIPPING CHARGES

United States

➤➤ Postal Book Rate { $3.00 First Item
50¢ Each Additional Item

➤➤ Priority Mail { $4.50 First Item
$2.00 Each Additional Item

➤➤ UPS { $5.00 First Item
$1.50 Each Additional Item

NOTE: UPS Delivery Available to Mainland USA Only

Canada

➤➤ Postal Book Rate { $6.00 First Item
$2.00 Each Additional Item

➤➤ Postal Air Mail { $8.00 First Item
$2.50 Each Additional Item

➤➤ Personal Checks or Bank Drafts MUST BE

USD$ and Drawn on a US Bank

➤➤ Canadian Postal Money Orders OK

➤➤ Payment MUST BE USD$

All Other Countries

➤➤ Surface Delivery { $10.00 First Item
$4.00 Each Additional Item

➤➤ Postal Air Mail { $14.00 First Item
$5.00 Each Additional Item

➤➤ Payment MUST BE USD$

➤➤ Checks and Money Orders MUST BE USD$
and Drawn on a US Bank or branch.

➤➤ Add $5.00 for Air Mail Subscription to
Future Adventures Unlimited Catalogs

SPECIAL NOTES

➤➤ RETAILERS: Standard Discounts Available

➤➤ BACKORDERS: We Backorder all Out-of-

Stock Items Unless Otherwise Requested

➤➤ PRO FORMA INVOICES: Available on Request

➤➤ VIDEOS: NTSC Mode Only. Replacement only.

➤➤ For PAL mode videos contact our other offices:

Please check: ☑

☐ This is my first order   ☐ I have ordered before

Name			
Address			
City			
State/Province		Postal Code	
Country			
Phone day		Evening	
Fax			

Item Code	Item Description	Qty	Total

Please check: ☑

Subtotal ➡
Less Discount-10% for 3 or more items ➡

☐ Postal-Surface   Balance ➡

☐ Postal-Air Mail Illinois Residents 6.25% Sales Tax ➡
(Priority in USA)   Previous Credit ➡

☐ UPS   Shipping ➡
(Mainland USA only) Total (check/MO in USD$ only) ➡

☐ Visa/MasterCard/Discover/Amex

Card Number

Expiration Date

10% Discount When You Order 3 or More Items!